Sewing in 1776.

Sewing in 1876.

SMITHSONIAN INSTITUTION

UNITED STATES NATIONAL MUSEUM

BULLETIN 254

WASHINGTON, D.C.

1968

THE SMITHSONIAN INSTITUTION PRESS

The Invention
Of the Sewing Machine

Grace Rogers Cooper

CURATOR OF TEXTILES
MUSEUM OF HISTORY AND TECHNOLOGY

SMITHSONIAN INSTITUTION • WASHINGTON, • D.C. 1968

Publications of the United States National Museum

The scholarly and scientific publications of the United States National Museum include two series, *Proceedings of the United States National Museum* and *United States National Museum Bulletin.*

In these series, the Museum publishes original articles and monographs dealing with the collections and work of its constituent museums—The Museum of Natural History and the Museum of History and Technology—setting forth newly acquired facts in the fields of anthropology, biology, history, geology, and technology. Copies of each publication are distributed to libraries, to cultural and scientific organizations, and to specialists and others interested in the different subjects.

The *Proceedings*, begun in 1878, are intended for the publication, in separate form, of shorter papers from the Museum of Natural History. These are gathered in volumes, octavo in size, with the publication date of each paper recorded in the table of contents of the volume.

In the *Bulletin* series, the first of which was issued in 1875, appear longer, separate publications consisting of monographs (occasionally in several parts) and volumes in which are collected works on related subjects. *Bulletins* are either octavo or quarto in size, depending on the needs of the presentation. Since 1902 papers relating to the botanical collections of the Museum of Natural History have been published in the *Bulletin* series under the heading *Contributions from the United States National Herbarium* and, since 1959, in *Bulletins* titled "Contributions from the Museum of History and Technology," have been gathered shorter papers relating to the collections and research of that Museum.

This work forms volume 254 of the *Bulletin* series.

FRANK A. TAYLOR
Director, United States National Museum

Contents

Preface

It had no instrument panel with push-button controls. It was not operated electronically or jet-propelled. But to many 19th-century people the sewing machine was probably as awe-inspiring as a space capsule is to their 20th-century descendants. It was expensive, but, considering the work it could do and the time it could save, the cost was more than justified. The sewing machine became the first widely advertised consumer appliance, pioneered installment buying and patent pooling, and revolutionized the ready-made clothing industry. It also weathered the protests of those who feared the new machine was a threat to their livelihood.

The practical sewing machine is not the result of one man's genius, but rather the culmination of a century of thought, work, trials, failures, and partial successes of a long list of inventors. History is too quick to credit one or two men for an important invention and to forget the work that preceded and prodded each man to contribute his share. It is no discredit to Howe to state that he *did not invent the sewing machine*. Howe's work with the sewing machine was important, and he did patent certain improvements, but his work was one step along the way. It is for the reader to decide whether it was the turning point.

Since the sewing machine has been considered by some as one of the most important inventions of 19th-century America, of equal importance to this story of the invention is the history of the sewing machine's development into a practical, popular commodity. Since many new companies blossomed overnight to manufacture this very salable item, a catalog list of more than one hundred and fifty of these 19th-century companies is included in this study. Still, the list is probably incomplete. Many of the companies remained in business a very short time or kept their activities a secret to avoid payment of royalties to patent holders. Evidence of these companies is difficult to find. It is hoped that additional information will come to light as a result of this initial attempt to list and date known companies. The dating of individual machines based on their serial numbers is also a difficult task. Individual company records of this type have not survived; however, using the commercial machines in the patent collection, for which we know one limiting date—the date the machine was deposited at the patent office—and using the records that have survived, an estimated date based on the serial number can be established for many of the better known machines.

Acknowledgments

I am greatly indebted to the late Dr. Frederick Lewton, whose interest in the history of the sewing machine initiated the collecting of information about it for the Smithsonian Institution's Division of Textiles archives and whose out-of-print booklet "A Servant in the House" prompted the writing of this work. I would also like to thank Mr. Bogart Thompson of the Singer Manufacturing Company for his cooperation in arranging for the gift of an excellent collection of 19th-century sewing machines to the Smithsonian and for allowing me to use the Singer historical files. Acknowledgment is also made of the cooperation extended by The Henry Ford Museum and Greenfield Village for permitting me to study their collection of old sewing machines.

Grace Rogers Cooper

Chapter One

Figure 1.—AFTER ALMOST A CENTURY OF ATTEMPTS TO INVENT A MACHINE THAT WOULD SEW, the practical sewing machine evolved in the mid-19th century. This elegant, carpeted salesroom of the 1870s, with fashionable ladies and gentlemen scanning the latest model sewing machines, reflects the pinnacle reached by the new industry in just a few decades. This example, one of many of its type, is the Wheeler and Wilson sewing-machine offices and salesroom, No. 44 Fourteenth Street, Union Square, New York City. From *The Daily Graphic*, New York City, December 29, 1874. (Smithsonian photo 48091–A.)

Early Efforts

To 1800

FOR THOUSANDS OF YEARS, the only means of stitching two pieces of fabric together had been with a common needle and a length of thread. The thread might be of silk, flax, wool, sinew, or other fibrous material. The needle, whether of bone, silver, bronze, steel, or some other metal, was always the same in design—a thin shaft with a point at one end and a hole or eye for receiving the thread at the other end. Simple as it was, the common needle (fig. 2) with its thread-carrying eye had been an ingenious improvement over the sharp bone, stick, or other object used to pierce a hole through which a lacing then had to be passed.[1] In addition to utilitarian stitching for such things as the making of garments and household furnishings, the needle was also used for decorative stitching, commonly called embroidery. And it was for this purpose that the needle, the seemingly perfect tool that defied improvement, was first altered for ease of stitching and to increase production.

One of the forms that the needle took in the process of adaptation was that of the fine steel hook. Called an *aguja* in Spain, the hook was used in making a type of lace known as *punto de aguja*. During the 17th century after the introduction of chainstitch embroideries from India, this hook was used to produce chainstitch designs on a net ground.[2] The stitch and the fine hook to make it were especially adaptable to this work. By the 18th century the hook had been reduced to needle size and inserted into a handle, and was used to chainstitch-embroider woven fabrics.[3] In France the hook was called a crochet and was sharpened to a point for easy entry into the fabric (fig. 3). For stitching, the fabric was held taut on a drum-shaped frame. The hooked needle pierced the fabric, caught the thread from below the surface and pulled a loop to the top. The needle reentered the fabric a stitch-length from the first entry and caught the thread again, pulling a second loop through the first to which it became enchained. This method of embroidery permitted for the first time the use of a continuous length of thread. At this time the chainstitch was used exclusively for decorative embroidery, and from the French name for drum—the shape of the frame that held the fabric—the worked fabric came to be called tambour embroidery. The crochet [4]

---

[1] CHARLES M. KARCH, *Needles: Historical and Descriptive* (12 Census U.S., vol. X, 1902), pp. 429–432.

[2] FLORENCE LEWIS MAY, *Hispanic Lace and Lace Making* (New York, 1939), pp. 267–271.

[3] Diderot's *L'Encyclopédie, ou dictionnaire raisonné des sciences, des arts et des métiers* . . . , vol. II (1763), Plates Brodeur, plate II.

[4] The term "crochet," as used today, became the modern counterpart of the Spanish *punto de aguja* about the second quarter of the 19th century.

or small hooked needle soon became known as a tambour needle.

In 1755 a new type of needle was invented for producing embroidery stitches. This needle had to pass completely through the fabric two times (a through-and-through motion) for every stitch. The inventor was Charles F. Weisenthal, a German mechanic living in London who was granted British patent 701 for a two-pointed needle (fig. 4). The invention was described in the patent as follows:

> The muslin, being put into a frame, is to be worked with a needle that has two points, one at the head, and the other point as a common needle, which is to be worked by holding it with the fingers in the middle, so as not to require turning.

It might be argued that Weisenthal had invented the eye-pointed needle, since he was the first inventor to put a point at the end of the needle having the eye. But, since his specifically stated use required the needle to have two points and to be passed completely through the fabric, Weisenthal had no intention of utilizing the very important advantage that the eye-pointed needle provided, that of *not* requiring the passage of the needle through the fabric as in hand sewing.

While no records can be found to establish that Weisenthal's patent was put to any commercial use during the inventor's lifetime, the two-pointed needle with eye at midpoint appeared in several 19th-century sewing-machine inventions.

The earliest of the known mechanical sewing devices produced a chain or tambour stitch, but by an entirely different principle than that used with either needle just described. Although the idea was incorporated into a patent, the machine was entirely overlooked for almost a century as the patent itself was classed under wearing apparel. It was entitled "An Entire New Method of Making and Completing Shoes, Boots, Splatterdashes, Clogs, and Other Articles, by Means of Tools and Machines also Invented by Me for that Purpose, and of Certain Compositions of the Nature of Japan or Varnish, which will be very advantageous in many useful Applications." This portentously titled British patent 1,764 was issued to an English cabinetmaker, Thomas Saint, on July 17, 1790. Along with accounts of several processes for making various varnish compositions, the patent contains descriptions of three separate machines; the second of these was for "stitching, quilting, or sewing." Though far from practical, the machine incorporated

Figure 2.—Primitive needle. Bronze. Egyptian (Roman period, 30 B.C.–A.D. 642). (Smithsonian photo 1379–A.)

several features common to a modern sewing machine. It had a horizontal cloth plate or table, an overhanging arm carrying a straight needle, and a continuous supply of thread from a spool. The motion was derived from the rotation of a hand crank on a shaft, which activated cams that produced all the actions of the machine.

One cam operated the forked needle (fig. 5) that pushed the thread through a hole made by a preceding thrust of the awl. The thread was caught by a looper and detained so that it then became enchained in the next loop of thread. The patent described thread tighteners above and below the work and an adjustment to vary the stitches for different kinds of material. Other than the British patent records, no contemporary reference to Saint's machine has ever been found. The stitching-machine contents of this patent was happened on by accident in 1873.[5] Using the patent description, a Newton Wilson of London attempted to build a model of Saint's machine in 1874.[6] Wilson found, however, that it was necessary to modify the construction before the machine would stitch at all.

---

[5] *Sewing Machine News* (1880), vol. 1, no. 7, p. 2.

[6] This model of Saint's machine was bequeathed by Mr. Wilson to the South Kensington Museum, London, England.

4

Figure 3.—Tambour needle and frame, showing the method of forming the chainstitch, from the Diderot Encyclopedia of 1763, vol. II, *Plates Brodeur*, plate II. (Smithsonian photo 43995–C.)

This raised the question whether Saint had built even one machine. Nevertheless, the germ of an idea was there, and had the inventor followed through the sewing machine might have been classed an 18th-century rather than a 19th-century contribution.

Figure 4.—WEISENTHAL'S two-pointed needle, 1755.

1800–1820

There is no doubt that the successful late-18th-century improvements in spinning and weaving methods, resulting in increased production of fabrics, had a great effect in spurring inventors to ideas of stitching by machinery. Several efforts were made during the first two decades of the 19th century to produce such machines.

On February 14, 1804, a French patent was issued to Thomas Stone and James Henderson for a "new mechanical principle designed to replace handwork in joining the edges of all kinds of flexible material, and particularly applicable to the manufacture of clothing." [7] The machine used a common needle and made an overcast stitch in the same manner as hand sewing. A pair of jaws or pincers, imitating the action of the fingers, alternately seized and released the needle on each side of the fabric. The pincers were attached to a pair of arms arranged to be moved backward and forward by "any suitable mechanism." [8] This machine was capable of making curved or angular as well as straight seams, but it was limited to carrying a short length of thread, necessitating frequent rethreading. The machine may have had some limited use, but it was not commercially successful.

On May 30 of the same year John Duncan, a Glasgow manufacturer, was granted British patent 2,769 for "a new and improved method of tambouring, or raising flowers, figures or other ornaments upon muslins, lawns and other cottons, cloths, or stuffs." This machine made the chainstitch, using not one but many hooked needles that operated simultaneously. The needles, attached to a bar or carrier, were pushed through the vertically held fabric from the upper right side, which in this case was also the outer side. After passing through it, they were supplied with thread from spools by means of peculiarly formed hooks or thread carriers. The thread was twisted around the needle above the hook, so as to be caught by it, and drawn through to the outer surface. The shaft of the needle was grooved on the hook side and fitted with a slider. This slider closed upon the retraction of the needle from the fabric, holding the thread in place and preventing the hook from catching. The fabric was stretched between two rollers set in an upright frame capable of sliding vertically in a second frame arranged to have longitudinal motion. The combination of these two motions was sufficient to produce any required design. The principle developed by Duncan was used on embroidery machines, in a modified form, for many years. Of several early attempts, his was the first to realize any form of success.

A type of rope-stitching machine, which might be

Figure 5.—SAINT'S SEWING MACHINE, 1790. (Smithsonian photo 42490–A.)

[7] *Sewing Machine News* (1880), vol. 1, no. 8, p. 2.
[8] Ibid.

Figure 6.—CHAPMAN'S SEWING MACHINE, first eye-pointed needle, 1807. (Smithsonian photo 33299–K.)

considered unimportant to this study, must be included because of its use of the eye-pointed needle, the needle that was to play a most important part in the later development of a practical sewing machine. The earliest reference to the use of a needle with an eye not being required to be passed completely through the fabric it was stitching is found in a machine invented by Edward Walter Chapman, for which he and William Chapman were granted British patent 3,078 on October 30, 1807. The machine (fig. 6) was designed to construct belting or flat banding by stitching together several strands of rope that had been laid side by side. Two needles were required and used alternately. One needle was threaded and then forced through the ropes. On the opposite side the thread was removed from the eye of the first needle before it was withdrawn. The second needle was threaded and the operation repeated. The needles could also be used to draw the thread, rather than push it, through the ropes with the same result. While being stitched, the ropes were held fast and the sewing frame and supporting carriage were moved manually as each stitch was made. Such a machine would be applicable only to the work described, since the necessity of rethreading at every stitch would make it impractical for any other type of sewing.

Another early machine reported to have used the eye-pointed needle to form the chainstitch was invented about 1810 by Balthasar Krems,[9] a hosiery worker of Mayen, Germany. One knitted article produced there was a peaked cap, and Krems' machine was devised to stitch the turned edges of the cap,[10] which was suspended from wire pins on a moving wheel. The needle of the machine was attached to a horizontal shaft and carried the thread through the fabric. The loop of thread was retained by a hook-shaped pin to become enchained with the next loop at the reentry of the needle. Local history reports that this device may have been used as early as 1800, but the inventor did not patent his machine and apparently made no attempt to commercialize it. No contemporary references to the

[9] ERICH LUTH, *Ein Mayener Strumpfwirker, Balthasar Krems, 1760–1813, Erfinder der Nähmaschine*, p. 10, states that the machine used an eye-pointed needle. WILHELM RENTERS, *Praktisches wissen von der Nähmaschine*, p. 4, states that Krems used a hooked needle. Renters probably mistook the hooked retaining pin for the needle.

[10] Dr. Dahmen, Burgermeister of Mayen, stated in a letter of October 8, 1963, that the original Krems machine was turned over to the officials of Mayen by Krems' descendants about the turn of the century. He verified that the machine used an eye-pointed needle. About 1920 the machine was placed in the Eifelmuseum in Genovevaburg; some of the unessential parts were restored. The machine now at this museum is the one pictured in Luth's book. A replica of the machine is in the Deutsches Museum, Munich, Germany.

7

Figure 7.—Madersperger's 1814 sewing machine. Illustration from a pamphlet by the inventor entitled *Beschreibung einer Nähmaschine*, Vienna, ca. 1816. (Smithsonian photo 49373.)

machine could be found, and use of the machine may have died with the inventor in 1813.

About the same time, Josef Madersperger, a tailor in Vienna, Austria, invented a sewing machine, which was illustrated (fig. 7) and described in a 15-page pamphlet published about 1816.[11] On May 12, 1817, a Vienna newspaper wrote of the Madersperger machine: "The approbation which his machine

received everywhere has induced his Royal Imperial Majesty, in the year 1814, to give to the inventor an exclusive privilege [patent] which has already been mentioned before in these papers." [12] Madersperger's 1814 machine stitched straight or curving lines. His second machine stitched small semicircles, as shown in the illustration, and also small circles, egg-shaped figures, and angles of various degrees. The machine, acclaimed by the art experts, must therefore have been intended for embroidery stitching. From the contemporary descriptions and the illustration, the machine is judged to have made a couched stitch—one thread was laid on the surface of the fabric and stitched in place with a short thread carried by a two-pointed needle of the type invented by Weisenthal. Two fabrics could have been stitched together, but not in the manner required for tailoring. The machine must have had many deficiencies in the tension adjustment, feed, and related mechanical operations, for despite the published wishes for success the inventor did not put the machine into practical

[11] Josef Madersperger, *Beschreibung einer Nähmaschine* (Vienna, ca. 1816). The exact date of this small booklet is not known. In the booklet Madersperger reports that he had received a patent in 1814 for his *first* machine adapted to straight sewing. However, the machine described and illustrated in this booklet was one that could stitch semicircles and small figures. In *Kunst und Gewerbeblatt*, a periodical (Munich, Germany, 1817, pp. 336–338), reference is made to the Madersperger machine and a statement to the effect that the inventor had published a leaflet describing his machine. The leaflet referred to is believed to be the one under discussion. For this reason it must have been published between 1814 and 1817, therefore ca. 1816. The only copy of this booklet known to this author is in the New York Public Library. It was probably not known to authors Luth and Renters. The author wishes to thank Miss Rita J. Adrosko of her staff for her important help in translating these German publications.

[12] *Sewing Machine Times* (1907), vol. 26, no. 865, p. 1.

Figure 8.—An engraving of Thimonnier and his sewing machine of 1830, from *Sewing Machine News*, 1880. (Smithsonian photo 10569-C.)

operation.[13] Years later Madersperger again attempted to invent a sewing machine using a different stitch (see p. 13).

A story persists that about 1818–1819 a machine that formed a backstitch, identical to the one used in hand sewing, was invented in Monkton, Vermont. The earliest record of this machine that this author has found was in the second or 1867 edition of *Eighty Years of Progress of the United States;* the machine is

not mentioned in the earlier edition. The writer of the article on sewing machines states that John Knowles invented and constructed a sewing machine, which used a single thread and a two-pointed needle with the eye in the middle to form the backstitch. This information must have come to light after the first edition was published, but from where and by whom is not known. Other sources state that two men, Adams and Dodge, produced this machine in Monkton.[14] While still others credit the Reverend John Adam Dodge, assisted by a mechanic by the name of John Knowles, with the same invention in the same location.[15] Vermont historical societies have been unable to identify the men named or to verify the story of the invention.[16] The importance of the credibility of this story, if proved, rests in the fact that it represents the first effort in the United States to produce a mechanical stitching device.

1820–1845

American records of this period are incomplete as a result of the Patent Office fire of 1836, in which most of the specific descriptions of patents issued to that date were destroyed. Patentees were asked to provide another description of their patents so that these might be copied, but comparatively few responded and only a small percentage was restored. Thus, although the printed index of patents [17] lists Henry Lye as patenting a machine for "sewing leather, and so forth" on March 10, 1826, no description of the machine has ever been located. Many patents whose original claim was for only a mechanical awl to pierce holes in leather or a clamp to hold leather for hand stitching were claimed as sewing devices once a practical machine had evolved. But no evidence has ever been found that any of these machines performed the actual stitching operation.

The first man known to have put a mechanical

---

[13] There are no known models of these early Madersperger machines in existence. Although the *Sewing Machine Times* reported in the 1907 issue that the 1814 sewing machine was then on exhibition in the Museum of the Vienna Polytechnic, the illustration shown was of Madersperger's 1839 machine. In a letter from the director of the Technisches Museum für Industrie und Gewerbe in Vienna, received in 1962, it was stated that the original 1814 Madersperger machine was in their museum. The photographs that were sent, however, were of the 1839 machine. This machine is entirely different from the 1814–1817 machine, as can readily be seen by the reader (figs. 7 and 10).

[14] John P. Stambaugh, *A History of the Sewing Machine* (Hartford, Conn., 1872), p. 13; *Sewing Machine News* (July 1880), vol. 1, no. 12, p. 4.

[15] "Sewing Machines," *Johnson's Universal Cyclopaedia* (New York, 1878), vol. 4, p. 205. The 1874 edition does not include this reference to Rev. John Adam Dodge.

[16] Letters to the author from the Vermont Historical Society (Nov. 13, 1953) and the Bennington Historical Museum and Art Gallery (May 2, 1953).

[17] Edmund Burke. Commissioner of Patents, *List of Patents for Inventions and Designs Issued by the United States from 1790 to 1847* (Washington, 1847).

Figure 9.—An adapted drawing of Hunt's sewing machine published by the *Sewing Machine News*, vol. 2, no. 8, 1881, to give some idea of its construction and operation. "The frame of the machine (A) rested on a base (B) that was supported by a table. The wheel (C) worked on a central shaft (E) and was set in motion by hand or foot power. On the front of the wheel (C) was a raised cam (D) into which the connecting rod (F) engaged to communicate motion to the vibrating arm (G) pivoted to the frame at (H) and carrying at the end (g) the curved needle (I). The take-up (J) served to tighten the thread after each stitch; it was connected to the vibrating arm by a rod (K). The cloth (L) was held in a vertical position between the fingers or nippers (M), which were attached to the frame. The bar (N) was toothed on one side (n) to mesh with the geared wheel (o). The lever (P) was operated by a cam (m) upon the periphery of the wheel (C), and carried the vertical pawl (S) which meshed with the ratchet (T) and moved the cloth as each stitch was made. The shuttle (U) worked in its race (V); it was operated by the vibrating lever (W), the upper end of which engaged into a groove on the face of the wheel (C)." (Smithsonian photo 42554.)

sewing device into commercial operation was Barthelemy Thimonnier,[18] a French tailor. After several years of fruitless effort he invented a machine for which he received a French patent in 1830.[19] The machine (fig. 8) made a chainstitch by means of a barbed or hooked needle. The vertically held needle worked from an overhanging arm. The needle thrust through the fabric laid on the horizontal table, caught a thread from the thread carrier and looper beneath the table, and brought a loop to the surface of the fabric. When the process was repeated the second loop became enchained in the first. The needle was moved downward by the depression of a cord-connected foot treadle and was raised by the action of a spring. The fabric was fed through the stitching mechanism manually, and a regular rate of speed had to be maintained by the operator in order to produce stitches of equal length. A type of retractable thimble or presser foot was used to hold the fabric down as required.

The needle, and the entire machine, was basically an attempt to mechanize tambour embroidery, with which the inventor was quite familiar. Although this work, which served as the machine's inspiration, was always used for decorative embroidery, Thimonnier saw the possibilities of using the stitch for utilitarian purposes. By 1841 he had 80 machines stitching army clothing in a Paris shop. But a mob of tailors, fearing that the invention would rob them of a livelihood, broke into the shop and destroyed the machines. Thimonnier fled Paris, penniless. Four years later he had obtained new financial help, improved his machine to produce 200 stitches a minute, and organized the first French sewing-machine company.[20] The Revolution of 1848, however, brought this enterprise to an early end. Before new support could be found other inventors had appeared with better machines, and Thimonnier's was passed by. In addition to the two French patents Thimonnier also received a British patent with his associate Jean Marie Magnin in 1848 and one in the United States in 1850. He achieved no financial gain from either of these and died a poor man.

While Thimonnier was developing his chainstitch machine in France, Walter Hunt,[21] perhaps best described as a Yankee mechanical genius, was working on a different kind of sewing machine in the United States. Sometime between 1832 and 1834 he produced at his shop in New York a machine that made a lockstitch.[22] This stitch was the direct result of the mechanical method devised to produce the stitching and represented the first occasion an inventor had not attempted to reproduce a hand stitch. The lockstitch required two threads, one passing through a loop in the the other and both interlocking in the heart of the seam. At the time Hunt did not consider the sewing machine any more promising than several other inventions that he had in mind, and, after demonstrating that the machine would sew, he sold his interest in it for a small sum and did not bother to patent it.

A description—one of few ever published—and sketch of a rebuilt Hunt machine (fig. 9) appeared in an article in the *Sewing Machine News* in 1881.[23] The important element in the Hunt invention was an eye-pointed needle working in combination with a shuttle carrying a second thread. Future inventors were thus no longer hampered by the erroneous idea that the sewing machine must imitate the human hands and fingers. Though Hunt's machine stitched short, straight seams with speed and accuracy, it could not sew curved or angular work. Its stitching was not continuous, but had to be reset at the end of a short run. The validity of Hunt's claim as the inventor of the lockstitch and the prescribed method of making it was argued many times, especially during the Elias Howe patent suits of the 1850s. The decision against Hunt was not a question of invention,[24] but one of right to ownership or control. Hunt did little to promote his sewing machine and sold it together with the right to patent to George A. Arrowsmith.

---

[18] See Barthelemy Thimonnier's biographical sketch, p. 137.

[19] French patent issued to Barthelemy Thimonnier and M. Ferrand (who was a tutor at l'Ecole des Mines, Saint-Etienne, and helped finance the patent), July 17, 1830.

[20] The company was located at Villefranche-sur-Saône, but no name is recorded. See J. Granger, *Thimonnier et la machine à coudre* (1943), p. 16.

---

[21] See Walter Hunt's biographical sketch, p. 138.

[22] The earliest known reference in print to Walter Hunt's sewing machine is in *Sewing by Machinery: An Exposition of the History of Patentees of Various Sewing Machines and of the Rights of the Public* (I. M. Singer & Co., 1853). A more detailed story of Hunt's invention is in *Sewing Machine News* (1880–81), vol. 2, no. 2, p. 4; no. 4, p. 5; and no. 8, pp. 3 and 8.

[23] Vol. 2, no. 8, p. 3.

[24] In the opinion and decision of C. Mason, Commissioner of the Patent Office, offered on May 24, 1854, for the Hunt vs. Howe interference suit, Mason stated: "He [Hunt] proves that in 1834 or 1835 he contrived a machine by which he actually effected his purpose of sewing cloth with considerable success."

Figure 10.—Madersperger's 1839 sewing machine. Madersperger's machine consisted of two major parts: the frame, which held the material, and the stitching mechanism, called the hand. The hand shown here is an original model. (*Photo courtesy of Technisches Museum für Industrie und Gewerbe, Vienna.*)

For over fifteen years, from the mid-1830s to the early 1850s, the machine dropped out of sight. When the sewing-machine litigation developed in the 1850s, the I. M. Singer company searched out the Hunt machine, had the inventor rebuild one,[25] and attempted to use this to break the Howe patent. The plan did not work. The Honorable Charles Mason, Patent Commissioner, reported:

> When the first inventor allows his discovery to slumber for eighteen years, with no probability of its ever being brought into useful activity, and when it is only resurrected to supplant and strangle an invention which has been given to the public, and which has been made practically useful, all reasonable presumption should be in favor of the inventor who has been the means of conferring the real benefit upon the world.[26]

Hunt's machine was an invention of the 1830s, but only because of the patent litigation was it ever heard of again.

During the time that a potentially successful sewing machine was being invented and forgotten in America, Josef Madersperger of Austria made a second attempt to solve the mechanical stitching problem. In 1839 he received a second patent on a machine entirely different from his 1814 effort. It was similar to Hunt's in that it used an eye-pointed needle and passed a thread through the loop of the needle-thread—the thread carried by the needle—to lock the stitch. Madersperger's machine was a multiple-needle quilting machine. The threaded needles penetrated the fabric from below and were retracted, leaving the loops on the surface. A thread was drawn through the loops to produce what the inventor termed a chain. The first two stitches were twisted before insertion into the next two, producing a type of twisted lockstitch. The mechanism for feeding the cloth was faulty, however, and the inventor himself stated in the specifications that much remained to perfect and simplify it before its general application. (This machine was illustrated [fig. 10] in the *Sewing Machine Times*, October 25, 1907, and mistakenly referred to as the 1814 model.) Madersperger realized no financial gain from either venture and died in a poorhouse in 1850.

The first efforts of the 1840s reflected the work of the earlier years. In England, Edward Newton and Thomas Archbold invented and patented a machine on May 4, 1841, for tambouring or ornamenting the backs of gloves. Their machine used a hook on the upper surface to catch the loop of thread, but an eye-pointed needle from underneath was used to carry the thread up through the fabric. The machine was designed to use three needles for three rows of chainstitching, if required. Although the machine was capable of stitching two fabrics together, it was never contemplated as a sewing machine in the present use of the term. Their British patent 8,948 stated it was for "improvements in producing ornamental or tambour work in the manufacture of gloves."

The earliest American patent specifically recorded as a sewing machine was U.S. patent 2,466, issued to John J. Greenough on February 21, 1842. His machine was a short-thread model that made both the running stitch and the backstitch. It used the two-pointed needle, with eye at mid-length, which was passed back and forth through the material by means of a pair of pincers on each side of the seam. The pincers opened and closed automatically. The material to be sewn was held in clamps which moved it forward between the pincers to form a running stitch or moved it alternately backward and forward to produce a backstitch. The clamps were attached to a rack that automatically fed the material at a predetermined rate according to the length of stitch required. Since the machine was designed for leather or other hard material, the needle was preceded by an awl, which pierced a hole. The machine had a weight to draw out the thread and a stop-motion to stop the machinery when a thread broke or became too short. The needle was threaded with a short length of thread and required frequent refilling. Only straight seams could be stitched. The feed was continuous to the length of the rack bar; then it had to be reset. The motions were all obtained from the revolution of a crank. It is not believed that any machines, other than the patent model (fig. 11), were ever made. Little is known of Greenough other than his name.

In the succeeding year, on March 4, 1843, Benjamin W. Bean received the second American sewing-machine patent, U.S. patent 2,982. Like Greenough's, this machine made a running stitch, but by a different method. In Bean's machine the fabric was fed between the teeth of a series of gears. Held in a

---

[25] The rebuilt machine, according to a letter to the author from B. F. Thompson of the Singer company, is believed to have been one of the machines lost in a Singer factory fire at Elizabethport, N.J., in 1890.

[26] Op. cit. (footnote 24).

Figure 11.—Greenough's patent model, 1842. (Smithsonian photo 45525–G.)

groove in the gears was a peculiarly shaped needle bent in two places to permit it to be held in place by the gears and with a point at one end and the eye at the opposite end, as in a common hand needle. The action of the gears caused the fabric to be forced onto and through the threaded needle. Indefinite straight seams could be stitched as the fabric was continuously forced off the needle by the turning gears (fig. 12). A screw clamp held the machine to a table or other work surface. Machines of this and similar types reportedly had some limited usage in the dyeing and bleaching mills,[27] where lengths of fabric were stitched together before processing. Improved versions of Bean's machine were to be patented in subsequent years in England and America. The same principle was also used in home machines two decades later.

The third sewing-machine patent on record in the United States Patent Office is patent 3,389 issued on December 27, 1843, to George H. Corliss, better remembered as the inventor and manufacturer of the Corliss steam engine. It was his interest in the sewing machine, however, that eventually directed his attention to the steam engine.

Corliss had a general store at Greenwich, New York. A customer's complaint that the boots he had purchased split at the seams made Corliss wonder why someone had not invented a machine to sew stronger seams than hand-sewn ones. He considered the problem of sewing leather, analyzing the steps required to make the saddler's stitch, one popularly used in boots and shoes. He concluded that a sewing machine to do this type of work must first perforate the leather, then draw the threads through the holes, and finally secure the stitches by pulling the threads tight. The machine Corliss invented (fig. 13) was of the same general type as Greenough's, except that two two-pointed needles were required to make the saddler's stitch. This stitch was composed of two running stitches made simultaneously, one from each side.[28] The machine used two awls to pierce the holes through which the needles passed; finger levers

---

[27] EDWARD H. KNIGHT, *Sewing Machines*, vol. 3 of *Knight's American Mechanical Dictionary*.

---

[28] A seam using the saddler's stitch appears as a neat line of touching stitches on both sides. Even when made by hand, it is sometimes misidentified by the casual observer as the lockstitch because of the uniformity of both sides. If the saddler's stitch was formed of threads of two different colors, the even stitches on one side of the seam and the odd stitches on the reverse side would be of one color, and vice versa.

approached from opposite sides, seized the needles, pulled the threads firmly, and passed the needles through to repeat the operation. The working model that Corliss completed could unite two pieces of heavy leather at the rate of 20 stitches per minute.

Corliss, lacking capital, went to Providence, Rhode Island, in 1844 to secure backers. After months without success, he was forced to abandon the sewing machine and accept employment as a draftsman and designer. Though he considered himself a failure, this change of employment placed him on the threshold of his more rewarding life work, improvement of the steam engine.[29]

On July 22, 1844, James Rodgers was granted U.S. patent 3,672, the fourth American sewing-machine patent. The patent model is not known to be in existence, but this machine was of minor importance for it offered only a negligible change in the Bean running-stitch machine. The same corrugated gears were used but were placed in different positions so that one bend in the needle was eliminated. When Bean secured a reissue of his patent in 1849, he had adapted it to use a straight needle. Rodgers' machine is not known to have had any commercial success, although this type of machine experienced a brief period of popularity. By the early 1900s, however, the running-stitch machine was so little known that when one was illustrated in the *Sewing Machine Times* in 1907 [30] it excited more curiosity than any of the other early types.

Figure 12.—Bean's patent model, 1843. (Smithsonian photo 42490–C.)

On December 7, 1844, the same year that Rodgers secured his American patent, John Fisher and James Gibbons were granted British patent 10,424 for "certain improvements in the manufacture of figured or ornamental lace, or net, or other fabrics." From this superficial description of its work, the device might seem to be just another tambouring machine. It was not. Designed specifically for ornamental stitching, the machine made a two-thread stitch using an eye-pointed needle and a shuttle.[31] Several sets of needles and shuttles worked simultaneously. The needles were secured to a needlebar placed beneath the fabric. The shuttles were pointed at both ends to pass through each succeeding new loop formed by the needles. Each shuttle was activated by two vibrating arms worked by cams. Each needle was curved in the form of a bow, and in addition to the eye at the point each also had a second eye at the bottom of the curve. The shape of the needle

[29] *The Life and Works of George H. Corliss*, privately printed for Mary Corliss by the American Historical Society, 1930. The Corliss family records were turned over to the Baker Library, Harvard University. In a letter addressed to this author by Robert W. Lovett of the Manuscripts Division on August 2, 1954, it was reported that there was a record on their Corliss card to the effect that a model of his sewing machine, received with the collection, was turned over to the Massachusetts Institute of Technology; however, Mr. Lovett also stated that from a manuscript memoir of Mr. Corliss that it would seem that he developed only the one machine— the patent model. In a letter dated November 15, 1954, Stanley Backer, assistant professor of mechanical engineering, stated that after extensive inquiries they were unable to locate the model at M.I.T. In 1964, Dr. Robert Woodbury, of M.I.T., turned over to the Smithsonian Institution the official copies of the Corliss drawings and the specifications which had been awarded to the inventor by the Patent Office. It is possible that this may have been the material noted on the Harvard University card as having been transferred to M.I.T.

[30] *Sewing Machine Times* (July 10, 1907), vol. 26, no. 858, p. 1.

[31] This is the earliest known patent using the combination of an eye-pointed needle and a shuttle to form a stitch.

Figure 13.—CORLISS' PATENT MODEL, 1843. The piece of wood in the foreground is an enlarged model of the needle. (Smithsonian photo 42490.)

together with the position of the eyes permitted the pointed shuttle, carrying the second thread, to pass freely through the loop in the ascending needle thread. The fabric was carried by a pair of cloth rollers, capable of sliding in a horizontal plane in both a lateral and a lengthwise direction. These combined movements were sufficient to enable the operator to produce almost every embroidered design. The ornamenting, which might be a yarn, cord, or gimp, was carried by the shuttle thread. There was no tension on the shuttle thread, which was held in place by the thread from the needle. The stitch produced was a

form of couching.[32] It was in no sense a lockstitch. Fisher, who was the inventor, readily admitted at a later date that he had not had the slightest idea of producing a sewing machine, in the utilitarian meaning of the term. Although it has not been established that this machine was ever put into practical operation, Fisher's invention was to have a far-reaching effect on the development of the sewing machine in England.

[32] In embroidery, couching is the technique of laying a decorative thread on the surface of the fabric and stitching it into place with a second less-conspicuous thread.

Chapter Two

Figure 14.—HOWE'S PREPATENT MODEL of 1845, and the box used by the inventor to carry the machine to England in 1847. (Smithsonian photo 45506–B.)

Elements of a Successful Machine

THE REQUIREMENTS for producing a successful, practical sewing machine were a support for the cloth, a needle to carry the thread through the fabric and a combining device to form the stitch, a feeding mechanism to permit one stitch to follow another, tension controls to provide an even delivery of thread, and the related mechanism to insure the precise performance of each operation in its proper sequence. Weisenthal had added a point to the eye-end of the needle, Saint supported the fabric by placing it in a horizontal position with a needle entering vertically, Duncan successfully completed a chainstitch for embroidery purposes, Chapman used a needle with an eye at its point and did not pass it completely through the fabric, Krems stitched circular caps with an eye-pointed needle used with a hook to form a chainstitch, Thimmonier used the hooked needle to form a chainstitch on a fabric laid horizontally, and Hunt created a new stitch that was more readily adapted to sewing by machine than the hand stitches had been, but, although each may have had the germ of an idea, a successful machine had not evolved. There were to be hundreds of patents issued in an attempt to solve these and the numerous minor problems that would ensue. But the problems were solved. And, in spite of its Old World inception, the successful sewing machine can be credited as an American invention.

Although the invention of the practical sewing machine, like most important inventions, was a many-man project, historians generally give full credit to Elias Howe, Jr. Though such credit may be overly generous, Howe's important role in this history cannot be denied.

Elias Howe, Jr., was born on a farm near Spencer, Massachusetts, but he left home at an early age to learn the machinist's trade.[33] After serving an apprenticeship in Lowell, he moved to Boston. In the late 1830s, while employed in the instrument shop of Ari Davis, Howe is reported to have overheard a discussion concerning the need for a machine that would sew. In 1843, when illness kept him from his job for days at a time, he remembered the conversation and the promises of the rich reward that reputedly awaited the successful inventor. Determined to invent such a machine, he finally managed to produce sufficient results to interest George Fisher in buying a one-half interest in his proposed invention. By April 1845, Howe's machine (fig. 14) was used to sew all the seams of two woolen suits for men's clothing. He continued to demonstrate his machine but found that interest was, at best, indifferent.

Nevertheless, Howe completed a second machine (fig. 15), which he submitted with his application for a patent. The fifth United States patent (No. 4,750) for a sewing machine was issued to him on September 10, 1846. The machine used a grooved and curved eye-pointed needle carried by a vibrating arm, with the needle supplied with thread from a spool. Loops of thread from the needle were locked by a thread carried by a shuttle, which was moved

---

[33] See biographical sketch, pp. 138–141.

through the loop by means of reciprocating drivers. The cloth was suspended in a vertical position, impaled on pins projecting from a baster plate, which moved intermittently under the needle by means of a toothed wheel. The length of each stitching operation depended upon the length of the baster plate, and the seams were necessarily straight. When the end of the baster plate reached the position of the needle, the machine was stopped. The cloth was removed from the baster plate, which was moved back to its original position. The cloth was moved forward on the pins, and the seam continued.

In his patent specifications, Howe claimed the following:

1. The forming of the seam by carrying a thread through the cloth by means of a curved needle on the end of a vibrating arm, and the passing of a shuttle furnished with its bobbin, in the manner set forth, between the needle and the thread which it carried, under combination and arrangement of parts substantially the same with that described.

2. The lifting of the thread that passes through the needle-eye by means of the lifting-rod, for the purpose of forming a loop of loose thread that is to be subsequently drawn in by the passage of the shuttle, as herein fully described, said lifting-rod being furnished with a lifting pin, and governed in its motion by the guide-pieces and other devices, arranged and operating substantially as described.

3. The holding of the thread that is given out by the shuttle, so as to prevent its unwinding from the shuttle-bobbin after the shuttle has passed through the loop, said thread being held by means of the lever or slipping-piece, as herein made known, or in any other manner that is substantially the same in its operation and result.

4. The manner of arranging and combining the small lever with the sliding box, in combination with the spring-piece, for the purpose of tightening the stitch as the needle is retracted.

5. The holding of the cloth to be sewed by the use of a baster-plate furnished with points for that purpose, and with holes enabling it to operate as a rack in the manner set forth, thereby carrying the cloth forward and dispensing altogether with the necessity of basting the parts together.

The five claims, which were allowed Howe in his patent, have been quoted to show that he did not claim the invention of the eye-pointed needle, for which he has so often been credited. The court judgment [34]

Figure 15.—Howe's PATENT MODEL, 1846. (Smithsonian photo 45525–B.)

that upheld Howe's claim to his patented right to control the use of the eye-pointed needle in combination with a shuttle to form a lockstitch was mistakenly interpreted by some as verifying control of the eye-pointed needle itself.

After patenting his invention, Howe spent three discouraging years in both the United States and in England trying to interest manufacturers in building his sewing machine, under license. Finally, for £250 sterling, he sold the British patent rights to William Thomas and further agreed to adapt the machine to Thomas' manufacture of umbrellas and corsets.[35] This did not prove to be a financial success for Howe and by 1849 he was back in the United States, once again without funds.

On his return, Howe was surprised to find that other inventors were engaged in the sewing-machine

[34] In the Matter of the Application of Elias Howe, Jr. for an Extension of His Sewing Machine Patent Dated September 10, 1846, New York, 1860, with attachments A and B, U.S. Patent Office. [L.C. call no. TJ 1512.H6265]

[35] It is interesting to note that when William Thomas applied for the British patent of the Howe machine (issued Dec. 1, 1846), the courts would not allow the claim for the combination of the eye-pointed needle and shuttle to form a stitch, due to the Fisher and Gibbons patent of 1844. For more details on Howe's years in England see his biographical sketch, pp. 138–141.

Figure 16.—AN ENLARGEMENT of the stitching area. (Smithsonian photo 45525–B.)

problem and that sewing machines were being manufactured for sale. The sixth United States sewing-machine patent (No. 5,942) had been issued to John A. Bradshaw on November 28, 1848, for a machine specifically stated as correcting the defects in the E. Howe patent. Bradshaw did not purport that his machine was a new invention. His specifications read:

> The curved needle used in Howe's machine will not by itself form the loop in the thread, which is necessary for the flying bobbin, with its case, to pass through, and has, therefore, to be aided in that operation by a lifting-pin, with the necessary mechanism to operate it. This is a very bungling device, and is a great incumbrance to the action of the machine, being an impediment in the way of introducing the cloth to be sewed, difficult to keep properly adjusted, and very frequently gets entangled between the thread and the needle, by which the latter is frequently broken. This accident happens very often,

not withstanding all the precaution which it is possible for the most careful operator to exercise; and inasmuch as the delay occasioned thereby is very considerable, and the needles costly and difficult to replace, it is therefore very important that their breaking in this manner be prevented, which in my machine is done in the most effectual manner by dispensing with the lifting-pin altogether, the loop for the flying bobbin to pass through being made with certainty and of the proper form by means of my angular needle moved in a particular manner just before the flying-bobbin case is thrown. The shuttle and its bobbin for giving off the thread in Howe's machine are very defective . . . my neat and simple bobbin-case . . . gives off its thread with certainty and uniformity . . . . The baster-plate in the Howe machine is very inconvenient and troublesome . . . in my machine . . . the clamp . . . is a very simple and efficient device. . . . The Howe machine is stationary, and the baster-plate or cloth-holder progressive. The Bradshaw machine is progressive and the cloth-holder stationary.

21

Figure 17.—MOREY AND JOHNSON sewing machine, 1849. Below: The machine is marked with the name of its maker, Safford & Williams. The number 49 is a serial number. Missing parts have been replaced with plastic. (Smithsonian photo 48400; brass plate: 48400–H.)

Bradshaw's patent accurately described some of the defects of the Howe machine, but other inventors were later to offer better solutions to the problems.

Although the Bradshaw machine was not in current manufacture, a machine based on it received the seventh United States sewing-machine patent. Patent 6,099 was issued to Charles Morey and Joseph B. Johnson on February 6, 1849. Their machine (fig. 17) was being offered for sale even before the patent was issued.

This was the first American patent for a chainstitch machine. The stitch was made by an eye-pointed needle carrying the thread through the fabric; the thread was detained by a hook until the loop was enchained by the succeeding one. The fabric was held vertically by a baster plate in a manner similar to the Howe machine. Although not claimed in the patent description, the Morey and Johnson machine also had a bar device for stripping the cloth from the needle. This bar had a slight motion causing a yielding pressure to be exerted on the cloth. Although the patent was not granted until February 6, 1849, the application had been filed in April of the previous year. The machine was featured in the *Scientific American* on January 27, 1849 (fig. 18):

> Morey and Johnson Machine—These machines are very accurately adjusted in all their parts to work in harmony, without this they would be of no use. But they are now used in most of the Print Works and Bleach Works in New England, and especially by the East Boston Flour Company. It sews about one yard per minute, and we consider it superior to the London Sewing Machine the specification of which is in our possession. It [Morey and Johnson] is more simple—and this is a great deal. . . . The price of a machine and right to use $135. [36]

An improvement in the Morey and Johnson machine was patented by Jotham S. Conant for which he was issued a patent on May 8, 1849. Conant's machine offered a slight modification of the cloth bar and of the method of keeping the cloth taut during the stitching operation. No successful use of it is known.

A second improvement of the Morey and Johnson patent was also issued on May 8, 1849; this United States patent (No. 6,439) was to John Bachelder for the first continuous, but intermittent, sewing mechanism. As shown in the patent model (fig. 19), his clothholder consisted of an endless belt supported by and running around three or any other suitable number of cylindrical rollers. A series of pointed wires projected from the surface of the belt near the edge immediately adjacent to the needle. The wires could be placed at regular or irregular distances as required. The shaft of one of the cylindrical

---

[36] The machine referred to as the London Sewing Machine is the British patent of the Thimonnier machine. This patent was applied for by Jean Marie Magnin and was published by *Newton's London Journal*, vol. 39, p. 317, as Magnin's invention.

Figure 18.—A MOREY AND JOHNSON sewing machine as illustrated in *Scientific American*, January 27, 1849. (Smithsonian photo 45771.)

Figure 19.—BACHELDER'S PATENT MODEL, 1849. (Smithsonian photo 45572).

rollers, which supported the endless clothholder, carried a ratchet wheel advanced by the action of a pawl connected to the end of the crankshaft by a small crankpin, whose position or distance from the axis of rotation of the shaft could be adjusted.

By this adjustment the extent of the vertical travel of the impelling pawl was regulated to control the length of the stitch. A spring catch kept the ratchet wheel in place at the end of each forward rotation of the wheel by the pawl. A roller placed over the endless belt at its middle roller pressed the cloth onto the wire points. A curved piece of metal was bent over and down upon the top of the belt so that the cloth, as it was sewed, was carried toward and against the piece by the belt. The cloth rose upon and over the piece and was separated from the points. When the machine was in motion the cloth was carried forward, passed under the needle, was stitched, and finally, passed the separator and off the belt. A vertically reciprocating, straight, eye-pointed needle, a horizontal supporting surface, and a yielding cloth presser were all used, but none were claimed as part of the patent. These were later specifically claimed in reissues of this patent. Bachelder's one specific claim, the endless feed belt, was not limited to belt feeding

only. As he explained in the patent, a revolving table or a cylinder might be substituted.

Bachelder did not manufacture machines, but his patent was sold in the mid-1850s to I. M. Singer.[37] It eventually became one of the most important patents to be contributed to the "Sewing-Machine Combination," a patent pool, which is discussed in more detail on pages 41 and 42.

While new ideas and inventors continued to provide the answers to some of the sewing-machine problems, Elias Howe began a series of patent suits to sustain the rights that he felt were his. Since his interest had never been in constructing machines for sale, it was absolutely essential for Howe to protect his royalty rights in order to realize any return from his patent. He was reported [38] to have supervised the construction of 14 sewing machines at a shop[39] on Gold Street in New York toward the close of 1850. Sworn contemporary testimony indicates that the machines were of no practical use.[40] Elias stated, in

---

[37] The exact date is not known; however, it was prior to 1856 as the patent was included in the sewing-machine patent pool formed that year.

[38] JAMES PARTON, *History of the Sewing Machine*, p. 12, (originally published in the *Atlantic Monthly*, May 1867), later reprinted by the Howe Machine Company as a separate.

[39] *Sewing Machine Times* (Feb. 25, 1907), vol. 17, no. 382, p. 1, "His [Bonata's] shop was on Gold Street, New York, near the Bartholf shop, where Howe was building some of his early machines."

[40] *Sewing Machine News*, vol. 3, no. 5, p. 5, Sept. 1881–Jan. 1882. "History of the Sewing Machine."

Figure 20.—BLODGETT & LEROW SEWING MACHINE, 1850, as manufactured by A. Bartholf, New York; the serial number of the machine is 19. At right, an original brass plate from the same type of machine with needle arm and presser foot and arm, serial number 119; the plate, however, does not fit the machine correctly. (Smithsonian photo 48440–D; brass plate: 48440–K.)

his application for his patent extension,[41] that he made only one machine in 1850–51. In 1852 he advertised[42] territorial rights and machines, but apparently did not realize any financial success until he sold a half interest in his patent to George Bliss in November 1852.[43] Bliss later began manufacturing machines that he initially sold as "Howe's Patent"; however, these machines were substantially different from the basic Howe machine.

On May 18, 1853, Elias Howe granted his first royalty license to Wheeler, Wilson & Company. Within a few months licenses were also granted to Grover & Baker; A. Bartholf; Nichols & Bliss; J. A. Lerow; Woolridge, Keene, and Moore; and A. B. Howe, the brother of Elias. These licenses granted the manufacturer the right to use any part of the Howe patent,[44] but it did not mean that the machines

were Elias Howe machines. When a royalty license was paid, the patent date and sometimes the name was stamped onto the machine. For this reason, these machines are sometimes mistakenly thought to be Elias Howe machines. They are not.

Howe was also prevented from manufacturing a practical machine unless he paid a royalty to other inventors. Three of the major manufacturers and Howe resolved their differences by forming the "Sewing Machine Combination." Although Howe did not enter the manufacturing competition for many years, he profited substantially from the royalty terms of the combination. In 1860, he applied for and received a seven-year extension on his patent.

There were Howe family machines for sale during this period, but these were the ones that Amasa Howe had been manufacturing since 1853. The machine was an excellent one and received the highest medal for sewing machines, together with many flattering testimonials, at the London International Exhibition in 1862. After the publication of this award the demand for (Amasa) Howe sewing

---

[41] Op. cit. (footnote 34).

[42] *New York Daily Tribune*, Jan. 15, 1852, p. 2.

[43] See Howe's biographical sketch, p. 141.

[44] Op. cit. (footnote 34). Attachments A and B are copies of Judge Sprague's decisions.

Figure 21.—BLODGETT & LEROW SEWING MACHINE, 1850, stamped with the legend "Goddard, Rice & Co., Makers, Worcester, Mass." and the serial number 37. Below: An original brass plate marked "No. 38"; this plate fits the machine perfectly. (Smithsonian photo 48440–E; brass plate: 48440–J.)

machines was greatly increased at home and abroad. Elias took this opportunity to gain entry into the manufacturing business by persuading Amasa to let him build a factory at Bridgeport, Connecticut, and manufacture the (Amasa) Howe machines. Two years passed before the factory was completed, and Amasa's agents were discouraged. The loss could have been regained, but the machines produced at Bridgeport were not of the quality of the earlier machines. Amasa attempted to rebuild the Bridgeport machines, but finally abandoned them and resumed manufacturing machines in New York under his own immediate supervision.[45] Elias formed his own company and continued to manufacture sewing machines. In 1867 he requested a second extension of his patent, but the request was refused. Elias Howe died in October of the same year.

Meanwhile, another important sewing machine of a different principle had also been patented in 1849. This was the machine of Sherburne C.

Blodgett, a tailor by trade, who was supported financially by John A. Lerow. United States patent 6,766 was issued to both men on October 2, 1849. In the patent, the machine was termed as "our new 'Rotary Sewing Machine'." The shuttle movement was continuous, revolving in a circle, rather than reciprocating as in the earlier machines. Automatic tension was initiated, restraining the slack thread from interference with the point of the needle.

The Blodgett and Lerow machine was built by several shops. One of the earliest was the shop of Orson C. Phelps on Harvard Place in Boston. Phelps took the Blodgett and Lerow machine to the sixth exhibition of the Massachusetts Charitable Mechanics Association in September 1850 and won a silver medal and this praise, "This machine performed admirably; it is an exceedingly ingenious and compact machine, able to perform tailor's sewing beautifully and thoroughly." [46] Although Phelps had

[45] *Sewing Machine Journal* (July 1887), pp. 93–94.

[46] *Report of the Sixth Exhibition of the Massachusetts Charitable Mechanics Association, in the City of Boston, September 1850* (Boston, 1850).

Figure 22.—WILSON'S PREPATENT MODEL for his reciprocating-shuttle machine, 1850.
(Smithsonian photo 45525–A.)

won the earliest known premium for a sewing machine, and although the machine was produced commercially to a considerable extent (figs. 20 and 21), one outstanding flaw in its operation could not be overlooked. As the shuttle passed around the six-inch circular shuttle race, it put a twist in the thread (or took one out if the direction was reversed) at each revolution. This caused a constant breaking of the thread, a condition that could not be rectified without changing the principle of operation. Such required changes were later to lead I. M. Singer, another well-known name, into the work of improving this machine.

Also exhibited at the same 1850 mechanics fair was the machine of Allen B. Wilson. Wilson's machine received only a bronze medal, but his inventive genius was to have a far greater effect on the development of the practical sewing machine than the work of Blodgett and Lerow. A. B. Wilson[47] was one of the ablest of the early inventors in the field of mechanical stitching, and probably the most original.

Wilson, a native of Willett, New York, was a young cabinetmaker at Adrian, Michigan, in 1847 when he first conceived of a machine that would sew. He was apparently unaware of parallel efforts by inventors in distant New England. After an illness, he moved to Pittsfield, Massachusetts, and pursued his idea in earnest. By November 1848 he had produced the basic drawings for a machine that would make a lockstitch. The needle, piercing the cloth, left a loop of thread below the seam. A shuttle carrying a second thread passed through the loop, and as the tension was adjusted a completed lockstitch was formed (fig. 22). Wilson's shuttle was pointed on both ends to form a stitch on both its forward and backward motion, a decided improvement over the shuttles of Hunt and Howe, which formed stitches in only one direction. After each stitch the cloth was advanced for the next stitch by a sliding bar against which the cloth was held by a stationary presser. While the needle was still in the cloth and holding it, the sliding bar returned for a fresh grip on the cloth.

Wilson made a second machine, on the same principle, and applied for a patent. He was approached by the owners of the Bradshaw 1848 patent, who

47 See biographical sketch, pp. 141–142.

Figure 23.—WILSON'S PATENT MODEL, 1850. (Smithsonian photo 45504–H.)

Figure 24.—WILSON'S PREPATENT MODEL for his rotary hook, 1851. (Smithsonian photo 45506–E.)

claimed control of the double-pointed shuttle. Although this claim was without justification, as can be seen by examining the Bradshaw patent specifications, Wilson did not have sufficient funds to fight the claim. In order to avoid a suit, he relinquished to A. P. Kline and Edward Lee, a one-half interest in his U.S. patent 7,776 which was issued on November 12, 1850 (fig. 23).

Inventor Wilson had been associated with Kline and Lee (E. Lee & Co.) for only a few months, when, on November 25, 1850, he agreed to sell his remaining interest to his partners for $2,000. He retained only limited rights for New Jersey and for Massachusetts. The sale was fruitless for the inventor, as no payment was ever made. How much money E. E. Lee & Co. realized from the Wilson machine is difficult to determine, but they ran numerous ads in the 1851 and 1852 issues of *Scientific American*. A typical one reads:

A. B. Wilson's Sewing Machine, justly allowed to be the cheapest and best now in use, patented November 12, 1850; can be seen on exhibition at 195 and 197 Broadway (formerly the Franklin House, Room 23, third floor) or to E. E. Lee & Co., Earle's Hotel. Rights for territory or machines can be had by applying to George R. Chittenden, Agent.[48]

Figure 25.—WILSON'S ROTARY-HOOK PATENT MODEL, 1851. (Smithsonian photo 45505–B.)

Another reads:

A. B. Wilson's Sewing Machine . . . the best and only practical sewing machine—not larger than a lady's work box—for the trifling sum of $35.[49]

Wilson severed relations with Lee and Kline in early 1851 shortly after meeting Nathaniel Wheeler, who was to become his partner in a happier, more

[48] *Scientific American* (Dec. 6, 1851), vol. 7, no. 12, p. 95.

[49] Ibid. (Sept. 20, 1851), vol. 7, no. 1, p. 7.

Figure 26.—WILSON'S stationary-bobbin patent model, 1852; a commercial machine was used since Wheeler, Wilson, Co. had begun manufacturing machines the previous year. (Smithsonian photo 45504–B.)

profitable enterprise involving the sewing machine.

Wilson, with his two partners, was occupying a room in the old Sun Building at 128 Fulton Street, when Wheeler, on a business trip to New York City, learned of the Wilson sewing machine. Wheeler examined the machine, saw its possibilities, and at once contracted with E. Lee & Co. to make 500 of them. At the same time he engaged Wilson to go with him to Watertown, Connecticut, to perfect the machine and supervise its manufacture. Meanwhile, Wilson had been working on a substitute for the shuttle. He showed his model of the device, which became known as the rotary hook, to Wheeler who was so convinced of its superiority that he decided to develop this new machine and leave Wilson's first machine to the others, who, by degrees, had become its owners.

Wilson now applied all his effort to improving the rotary hook, for which he received his second patent on August 12, 1851 (figs. 24 and 25). Wheeler, his two partners Warren and Woodruff, and Wilson now formed a new copartnership—Wheeler, Wilson, and Company. They began the manufacture of the machines under the patent, which combined the rotary hook and a reciprocating bobbin. The rotary hook extended or opened more widely the loop of the needle thread, while a reciprocating bobbin carried its thread through the extended loop. To avoid litigation which the reciprocating bobbin might have caused, Wilson contrived his third outstanding invention—the stationary bobbin. This was a feature of the first machine produced by the new company

in 1851, though the patent for the stationary bobbin was not issued until June 15, 1852 (fig. 26).

In all reciprocating-shuttle machines a certain loss of power is incurred in driving forward, stopping, and bringing back the shuttle at each stitch; also, the machines are rather noisy, owing to the striking of the driver against the shuttle at each stroke. These objections were removed by Wilson's rotary hook and stationary bobbin. The locking of the needle thread with the bobbin thread was accomplished, not by driving a shuttle through the loop of the needle thread, but by passing that loop under the bobbin. The driving shaft carried the circular rotary hook, one of the sewing machine's most beautiful contrivances. The success of the machine is indicated in an article that appeared in the June 1853 issue of *Scientific American*:

> There are 300 of these machines now in operation in various parts of the country, and the work which they can perform cannot be surpassed . . . . The time must soon come when every private family that has much sewing to do, will have one of these neat and perfect machines; indeed many private families have them now . . . . The price of one all complete is $125; every machine is made under the eye of the inventor at the company's machine shop, Watertown, Connecticut, so that every one is warranted . . . agreement between Mr. Howe and Messrs. Wheeler, Wilson & Co., so every customer will be perfectly protected . . . .[50]

---

[50] Ibid. (June 4, 1853), vol. 7, no. 38, p. 298

Figure 27.—WILSON's four-motion-feed patent model, 1854, is not known to be in existence; this is a commercial machine of the period. The plate is stamped "A. B. Wilson, Patented Aug. 12, 1851, Watertown, Conn., No. 1. . . ." (Smithsonian photo 45504.)

This agreement was important to sales, as Elias Howe was known to have sued purchasers of machines, as well as rival inventors and companies.

The business was on a substantial basis by October 1853, and a stock company was formed under the name of Wheeler & Wilson Manufacturing Company.[51] A little more than a year later, on December 19, 1854, Wilson's fourth important patent (U.S. patent 12,116)—for the four-motion cloth feed—was issued to him (fig. 27). In this development, the flat-toothed surface in contact with the cloth moved forward carrying the cloth with it; then it dropped a little, so as not to touch the cloth; next it moved backward; then in the fourth motion it pushed up against the cloth and was ready to repeat the forward movements. This simple and effective feed method is still used today, with only minor modifications, in almost every sewing machine. This feed with the rotary hook and the stationary circular-disk bobbin, completed the essential features of Wilson's machine. It was original and fundamentally different from all other machines of that time.

The resulting Wheeler and Wilson machine made a lockstitch by means of a curved eye-pointed needle carried by a vibrating arm projecting from a rock shaft connected by link and eccentric strap with an eccentric on the rotating hook shaft. This shaft had at its outer end the rotary hook, provided with

a point adapted to enter the loop of needle thread. As the hook rotated, it passed into and drew down the loop of needle-thread, which was held by means of a loop check, while the point of the hook entered a new loop. When the first loop was cast off—the face of the hook being beveled for that purpose—it was drawn upward by the action of the hook upon the loop through which it was then passing. During the rotation of the hook each loop was passed around a disk bobbin provided with the second thread and serving the part of the shuttle in other machines. The four-motion feed was actuated in this machine by means of a spring bar and a cam in conjunction with the mandrel.

From the beginning, Wheeler and Wilson had looked beyond the use of the sewing machine solely by manufacturers and had seen the demand for a light-running, lightweight machine for sewing in the home. Wilson's inventions lent themselves to this design, and Wheeler and Wilson led the way to the introduction of the machine as a home appliance. Other manufacturers followed.

When the stock company was formed, Mr. Wilson retired from active participation in the business at his own request. His health had not been good, and a nervous condition made it advisable for him to be freed from the responsibility of daily routine. During this period Wilson's inventive contributions to the sewing machine continued as noted, and in addition he worked on inventions concerning cotton picking and illuminating gases.

Wheeler and Wilson's foremost competitor in the

[51] J. D. VAN SLYCK, *New England Manufactures and Manufactories*, vol. 2, pp. 672–682.

29

early years of sewing-machine manufacture was the Singer Company, which overtook them by 1870 and finally absorbed the entire Wheeler and Wilson Manufacturing Company in 1905.

The founder of this most successful 19th-century company was Isaac Singer, a native of Pittstown, New York.[52] Successively a mechanic, an actor, and an inventor, Singer came to Boston in 1850 to promote his invention of a machine for carving printers' wooden type. He exhibited the carving machine in Orson Phelps' shop, where the Blodgett and Lerow machines were being manufactured.

Because the carving machine evoked but little interest, Singer turned his attention to the sewing machine as a device offering considerable opportunity for both improvement and financial reward. Phelps liked Singer's ideas and joined with George Zieber, the publisher who had been backing the carving-machine venture, to support Singer in the work of improving the sewing machine. His improvements in the Blodgett and Lerow machine included a table to hold the cloth horizontally rather than vertically (this had been used by Bachelder and Wilson also), a yielding vertical presser foot to hold the cloth down as the needle was drawn up, and a vertically reciprocating straight needle driven by a rotary, overhanging shaft.

The story of the invention and first trial of the machine was told by Singer in the course of a patent suit sometime later:

> I explained to them how the work was to be fed over the table and under the presser-foot, by a wheel, having short pins on its periphery, projecting through a slot in the table, so that the work would be automatically caught, fed and freed from the pins, in place of attaching and detaching the work to and from the baster plate by hand, as was necessary in the Blodgett machine.
>
> Phelps and Zieber were satisfied that it would work. I had no money. Zieber offered forty dollars to build a model machine. Phelps offered his best endeavors to carry out my plan and make the model in his shop; if successful we were to share equally. I worked at it day and night, sleeping but three or four hours a day out of the twenty-four, and eating generally but once a day, as I knew I must make it for the forty dollars or not get it at all.
>
> The machine was completed in eleven days. About nine o'clock in the evening we got the parts together and tried it; it did not sew; the workmen exhausted

with almost unremitting work, pronounced it a failure and left me one by one.

> Zieber held the lamp, and I continued to try the machine, but anxiety and incessant work had made me nervous and I could not get tight stitches. Sick at heart, about midnight, we started for our hotel. On the way we sat down on a pile of boards, and Zieber mentioned that the loose loops of thread were on the upper side of the cloth. It flashed upon me that we had forgot to adjust the tension on the needle thread. We went back, adjusted the tension, tried the machine, sewed five stitches perfectly and the thread snapped, but that was enough. At three o'clock the next day the machine was finished. I took it to New York and employed Mr. Charles M. Keller to patent it. It was used as a model in the application for the patent.[53]

The first machine was completed about the last of September 1850. The partners considered naming the machine the "Jenny Lind," after the Swedish soprano who was then the toast of America. It was reported [54] to have been advertised under that name when the machine was first placed on the market, but the name was soon changed to "Singer's Perpendicular Action Sewing Machine" or simply the "Singer Sewing Machine"—a name correctly anticipated to achieve a popularity of its own.

According to the contract made by the partners, the hurriedly built first machine was to be sent to the Patent Office with an application in the name of Singer and Phelps. An application was made between the end of September 1850 and March 14, 1851, as Singer refers to it briefly in the application formally filed on April 16, 1851, stating, "My present invention is of improvements on a machine heretofore invented by me and for which an application is now pending." [55]

---

[52] See his biographical sketch, pp. 142–143.

[53] CHESTER McNEIL, *A History of the Sewing Machine* in Union Sales Bulletin, vol. 3, Union Special Sewing Machine Co., Chicago, Illinois, pp. 83–85. 1903.

[54] *Sewing Machine Times* (Aug. 25, 1908), vol. 18, no. 418.

[55] Singer gives this limited description of the first machine, with detailed improvements for which he was then applying for a patent: "In my previous machine, to which reference has been made, the bobbin was carried by the needle-carrier, and hence the motion of the needle had to be equal to the length of thread required to form the loop, which was objectionable, as in many instances this range of motion was unnecessarily long for all other purposes . . . ." Quoted from U.S. patent 8,294 issued to Isaac M. Singer, Aug. 12, 1851. It should be noted that in some instances there was a considerable lapse of time from the date a patent application was made until the patent was issued. In this case the handwritten specifications were dated March 14, 1851, and the formal Patent Office receipt was dated April 16, 1851.

Figure 28.—SINGER'S PATENT MODEL, 1851; a commercial machine was used, bearing the serial number 22. (Smithsonian photo 45572–D.)

Figure 29.—SINGER'S PERPENDICULAR ACTION sewing machine, an engraving from *Illustrated News*, June 25, 1853, which states: "The sewing machine has, within the last two years acquired a wide celebrity, and established its character as one of the most efficient labor saving instruments ever introduced to public notice . . . . We must not forget to call attention to the fact that this instrument is peculiarly calculated for female operatives. They should never allow its use to be monopolized by men." (Smithsonian photo 48091–D.)

In late December 1850 Singer had bought Phelps' interest in the company. Whether the first application was later abandoned by Singer or whether it was rejected is not known,[56] but a patent on the first application was never issued. The final disposition of this first machine has remained a mystery.[57]

A few machines were manufactured in late 1850 and early 1851, and these attracted considerable attention; orders began to be received in advance of production. The pending patent application did not delay the manufacture, and a number of machines were sold before August 12, 1851, when the patent was granted. The patent model is shown in figure 28.[58] It made a lockstitch by means of a straight eye-pointed needle and a reciprocating shuttle. The patent claims, as quoted from the specifications, were as follows:

1. Giving to the shuttle an additional forward motion after it has been stopped to close the loop, as described, for the purpose of drawing the stitch tight, when such additional motion is given at and in combination with

[56] If a patent was not approved, for any reason, the records were placed in an "Abandoned File." In 1930 Congress authorized the disposal of the old "Abandoned Files," requiring them to be kept for twenty years only. There are no Singer Company records giving an account of the first patent application.

[57] Its whereabouts was unknown as early as 1908, as stated in the *Sewing Machine Times* (Aug. 25, 1908), vol. 18, no. 418. Models of abandoned patents frequently remained at the Patent Office. Approximately 76,000 models were ruined in a Patent Office fire in 1877. In 1908 over 3000 models of abandoned patents were sold at auction. Either incident could account for the machine's disappearance.

[58] The patent model of 8,294 is a machine that bears the serial number 22; it was manufactured before April 18, 1851, the date it was recorded as received by the Patent Office.

Figure 30.—I. M. Singer & Co. New York showroom of the mid-1850s, as illustrated in *Frank Leslie's Illustrated Newspaper*, August 29, 1857; only manufacturing machines are shown in this illustration. (Smithsonian photo 48091–B.)

the feed motion of the cloth in the reverse direction, and the final upward motion of the needle, as described, so that the two threads shall be drawn tight at the same time, as described.

2. Controlling the thread during the downward motion of the needle by the combination of a friction-pad to prevent the slack above the cloth, with the eye on the needle-carrier for drawing back the thread, for the purposes and in the manner substantially as described.

3. Placing the bobbin from which the needle is supplied with thread on an adjustable arm attached to the frame, substantially as described, when this is combined with the carrying of the said thread through an eye or guide attached to and moving with the needle-carrier, as

described, whereby any desired length of thread can be given for the formation of the loop without varying the range of motion of the needle, as described.

The feeding described in the Singer patent was "by the friction surface of a wheel, whose periphery is formed with very fine grooves, the edges of which are slightly serrated, against which the cloth is pressed by a spring plate or pad." Although claimed by the inventor in the handwritten specifications, it was not allowed as original.

The machines manufactured by the Singer company (fig. 29) were duplicates of the patent model. These machines were quite heavy and intended for

Figure 31.—Hunt and Webster's sewing-machine manufactory exhibition and salesroom in Boston, as illustrated in *Ballou's Pictorial*, July 5, 1856; only manufacturing machines are shown. (Smithsonian photo 45771–A.)

manufacturing rather than for family use in the home.

Singer enjoyed demonstrating the machine and showed it to church and social groups and even at circuses; this personal association then encouraged him to improve its reliability and convenience. He developed a wooden packing case which doubled as a stand for the machine and a treadle to allow it to be operated by foot. Because of the dimensions of the packing case, Singer put the pivot of the treadle toward its center, about where the instep of the foot would rest. This produced the heel-and-toe action treadle, a familiar part of the sewing machine until its replacement by the electric motor. Both hands were freed to guide and arrange the cloth that was being stitched. Singer also added a flywheel to smooth out the treadle action and later an iron stand with a treadle wide enough for both feet. The treadle had been in use for two years before a rival pointed out that it might have been patented. To Singer's chagrin it was then too late for patent laws

did not permit patenting a device that had been in public use.

A new obstacle appeared in the Singer company's path when Howe demanded $25,000 for infringement of his patent. Singer and Zieber decided to fight, enlisting the legal aid of Edward Clark, a lawyer and financier. Howe's action was opposed on the basis of Hunt's machine of 1834, which they stated had anticipated Howe's invention.

While they were resisting, Howe sued three firms that were using and selling Singer machines. The court order required the selling firms and the purchasers to provide an account of the profits accrued from the sale and the use of the sewing machines and restrained the firms from selling the machines during the pendency of the suit.[59] As a result of this action, a

---

[59] William R. Bagnall, in "Contributions to American Economic History," vol. 1 (1908), MS, Harvard School of Business Library.

Figure 33.—Singer Family Machine, 1858, head only. (Smithsonian photo 45524–F.)

Figure 32.—Singer's new Family Sewing Machine, illustration from a brochure dating about 1858 or 1859 which states: "A few months since, we came to the conclusion that the public taste demanded a sewing machine for family purposes more exclusively; a machine of smaller size, and of a lighter and more elegant form; a machine decorated in the best style of art, so as to make a beautiful ornament in the parlor or boudoir; a machine very easily operated, and rapid in working . . . . To supply this public want, we have just produced, and are now prepared to receive orders for, 'Singer's new Family Sewing Machine.'" (Smithsonian photo 48091–H.)

number of Singer's rivals purchased licenses from Howe and advertised that anyone could sell their machines without fear of a suit. This gave them a great competitive advantage, and Singer and Clark[60] decided it was best to seek a settlement with Howe. On July 1, 1854, they paid him $15,000 and took out a license.

In spite of this defeat, the Singer company could

---

[60] Singer purchased Phelps' interest in the company in 1851 and sold it to Edward Clark.

claim several important improvements to the sewing machine and the acquisition of the patents rights to the Morey and Johnson machine of 1849, which gave them control of the spring or curved arm to hold the cloth by a yielding pressure. Although this point had not been claimed in the 1849 patent, the established principle of patent law allowed that a novel device introduced and used in a patented machine could be covered by a reissue at any time during the life of the patent. Upon becoming owners of the Morey and Johnson patent, Singer applied for a reissue which covered this type of yielding pressure. It was granted on June 27, 1854. The Singer company's acquisition of the Bachelder patent had given them control of the yielding pressure bar also.

Singer's aggressive selling had begun to overcome the public's suspicion of sewing machines. He pioneered in the use of lavishly decorated sewing-machine showrooms when the company offices were expanded in the mid-1850s (fig. 30). These were rich with carved walnut furniture, gilded ornaments, and carpeted floors, places in which Victorian women were not ashamed to be seen. The machines were demonstrated by pretty young women. The total effect was a new concept of selling, and Singer became the drum major of a new and coming industry that had many followers (see fig. 31).

The first, light, family sewing machine by the Singer

Figure 34.—GROVER AND BAKER'S PATENT MODEL, 1851. (Smithsonian photo 32003–G.)

company was not manufactured until 1858 (figs. 32 and 33). Comparatively few of these machines were made as they proved to be too small and light. The men in the shop dubbed the machine "The Grasshopper," but it was officially called the new Family Sewing Machine or the Family Machine.[61] Because of its shape, Singer company brochures of the 1920s referred to it as the Turtleback Machine.

Since the cost of sewing machines was quite high and the average family income was low, Clark suggested the adoption of the hire-purchase plan. Into

[61] This first, family sewing machine should not be confused in name with a model brought out in the sixties. The name of this first, family machine was in the sense of a new "family" sewing machine. In 1859 a "Letter A" family machine was introduced. Thus in 1865 when the Singer Company brought out another family machine they called it the "New" Family Sewing Machine. Both the first-style Family machine and the Letter A machine are illustrated in *Eighty Years of Progress of the United States* (New York, 1861), vol. 2, p. 417, and discussed in an article, "The Place and Its Tenants," in the *Sewing Machine Times* (Dec. 25, 1908), vol. 27, no. 893.

the American economy thus came the now-familiar installment buying.

Singer and Clark continued to be partners until 1863 when a corporation was formed. At this time Singer decided to withdraw from active work. He received 40 percent of the stock and retired to Paris and later to England, where he died in 1875.

By the mid-1850s the basic elements of a successful, practical sewing machine were at hand, but the continuing court litigation over rival patent rights seemed destined to ruin the economics of the new industry. It was then that the lawyer of the Grover and Baker company, another sewing-machine manufacturer of the early 1850s, supplied the solution. Grover and Baker were manufacturing a machine that was mechanically good, for this early period. William O. Grover was another Boston tailor, who, unlike many others, was convinced that the sewing machine was going to revolutionize his chosen trade. Although the sewing machines that he had seen were not very practical, he began in 1849 to experiment with an idea based on a new kind of stitch. His design was for a machine that would take both its threads from

35

Figure 35.—This Grover and Baker cabinet-style sewing machine of 1856 bears the serial number 5675 and the patent dates February 11, 1851, June 22, 1852, February 22, 1853, and May 27, 1856. (Smithsonian photo 45572–F.)

spools and eliminate the need to wind one thread upon a bobbin. After much experimenting, he proved that it was possible to make a seam by interlocking two threads in a succession of slipknots, but he found that building a machine to do this was a much more difficult task. It is quite surprising that while he was working on this idea, he did not stumble upon a good method to produce the single-thread (as opposed to Grover and Baker's two-thread) chainstitch, later worked out by another. Grover was working so intently on the use of two threads that apparently no thought of forming a stitch with one thread had a chance to develop.

At this time Grover became a partner with another Boston tailor, William E. Baker, and on February 11, 1851, they were issued U.S. patent No. 7,931 for a machine that did exactly what Grover had set out to do; it made a double chainstitch with two threads both carried on ordinary thread spools. The machine (figs. 34 and 35) used a vertical eye-pointed needle for the top thread and a horizontal needle for the underthread. The cloth was placed on the horizontal platform or table, which had a hole for the entry of the vertical needle. When this needle passed through the cloth, it formed a loop on the underside. The horizontal needle passed through this loop forming another loop beyond, which was retained until the redescending vertical needle

Figure 36.—Grover's patent model for the first portable case, 1856. The machine in the case is a commercial machine of 1854, bearing the serial number 3012 and the patent dates "Feby 11, 1851, June 22, 1852, Feby 22, 1853." Powered by a single, foot-shaped treadle that was connected by a removable wooden pitman, it also could be turned by hand. (Smithsonian photo 45525–D.)

enchained it, and the process repeated. The slack in the needle thread was controlled by means of a spring guide. The cloth was fed by feeding rolls and a band.

A company was organized under the name of Grover and Baker Sewing Machine Company, and soon the partners took Jacob Weatherill, mechanic, and Orlando B. Potter, lawyer (who became the president), into the firm. Potter contributed his ability as a lawyer in lieu of a financial investment and handled the several succeeding patents of Grover and Baker. These patents were primarily for mechanical improvements such as U.S. patent No. 9,053 issued to Grover and Baker on June 22, 1852, for devising a curved upper needle and an under looper [62] to form the double-looped stitch which became known as the Grover and Baker stitch.

[62] A looper on the underside in place of the horizontal needle.

One of the more interesting of the patents, however, was for the box or sewing case for which Grover was issued U.S. patent No. 14,956 on May 27, 1856. The inventor stated "that when open the box shall constitute the bed for the machine to be operated upon, and hanging the machine thereto to facilitate oiling, cleansing, and repairs without removing it from the box." It was the first portable sewing machine (fig. 36).

Though the Grover and Baker company manufactured machines using a shuttle and producing the more common lockstitch, both under royalty in their own name and also for other smaller companies, Potter was convinced that the Grover and Baker stitch was the one that eventually would be used in both family and commercial machines. He, as president, directed the efforts of the company to that end. When the basic patents held by the "Sewing-Machine Combination" (discussed on pp. 41–42) began to run out in the mid-1870s, dissolving its purpose and lowering the selling price of sewing machines, the Grover and Baker company began a systematic curtailing of expenses and closing of branch offices. All the patents held by the company and the business itself were sold to another company.[63] But the members of the Grover and Baker company fared well financially by the strategic move.

The Grover and Baker machine and its unique stitch did not have a great influence on the overall development of the mechanics of machine sewing. The merits of a double-looped stitch—its elasticity and the taking of both threads from commercial spools—were outweighed by the bulkiness of the seam and its consumption of three times as much thread as the lockstitch required. Machines making a similar type of stitch have continued in limited use in the manufacture of knit goods and other products requiring an elastic seam. But, more importantly, Grover and Baker's astute Orlando B. Potter placed their names in the annals of sewing-machine history by his work in forming the "Combination," believed to be the first "trust" of any prominence.

---

[63] Domestic Sewing Machine Company. See *Union Special Sewing Machine Co. Sales Bulletin*, vol. 3, ch. 15, pp. 58–59.

Chapter Three

A PARTIAL STATEMENT FROM RECORDS OF "THE SEWING-MACHINE COMBINATION," SHOWING NUMBER OF SEWING-MACHINES LICENSED ANNUALLY UNDER THE *ELIAS HOWE* PATENT.

Name of Manufacturer.	1853.	1854.	1855.	1856.	1857.	1858.	1859.	1860.	1861.	1862.	1863.	1864.	1865.	1866.
Wheeler & Wilson Mfg. Co...	799	756	1,171	2,210	4,591	7,978	21,306	25,102	18,556	28,202	29,778	40,062	39,157	50,132
I. M. Singer & Co..........	810	879	883	2,564	3,630	3,594	10,953	13,000 (a)	16,000 (a)	18,396	....	....	....	....
The Singer Manufacturing Co.	...	...	....	....	....	....	....				20,030	23,632	26,340	30,960
Grover & Baker S. M. Co...	657	2,034	1,144	1,952	3,680	5,070	10,280	(b)	(b)	(b)	(b)	(b)	(b)	(b)
A. B. Howe " " ..	...	60	53	47	133	179	921	(b)	(b)	(b)	(b)	(b)	(b)	(b)
Leavitt " " ..	28	217	152	235	195	75	213	(b)	(b)	(b)	(b)	(b)	(b)	(b)
Ladd & Webster " " ..	100	268	73	180	453	490	1,788	(b)	(b)	(b)	(b)	(b)	(b)	(b)
Bartholf " " ..	135	55	31	35	31	203	747	(b)	(b)	(b)	(b)	(b)	(b)	(b)

A PARTIAL STATEMENT SHOWING NUMBER OF SEWING-MACHINES LICENSED ANNUALLY FROM 1867 TO 1876 INCLUSIVE.

Name of Manufacturer.	1867.	1868.	1869.	1870.	1871.	1872.	1873.	1874.	1875.	1876.
The Singer Manufacturing Company...	43,053	59,629	86,781	127,833	181,260	219,758	232,444	241,679	249,852	262,316
Wheeler & Wilson Mfg. Company......	38,055	(b)	78,866	83,208	128,526	174,088	119,190	92,827	103,740	108,997
Grover & Baker Sewing-Machine Co...	32,999	35,000 (a)	35,188	57,402	50,838	52,010	36,179	20,000 (a)	15,000 (a)	....
Weed Sewing-Machine Co.....	3,638	12,000	19,687	35,002	39,655	42,444	21,769	20,495	21,993	14,425
Howe Sewing-Machine Co............	11,053	35,000 (a)	45,000 (a)	75,156	134,010	145,000 (a)	90,000 (a)	35,000 (a)	25,000 (a)	109,294
A. B. Howe " "					20,051					
B. P. Howe " "						14,907	13,919			
Willcox & Gibbs Sewing-Machine Co...	14,152	15,000	17,201	28,890	30,127	33,639	15,881	13,710	14,522	12,758
Wilson (W. G.) " "					21,153	22,666	21,247	17,525	9,508	
American B. H. & S. M. Co..........			7,792	14,573	20,121	18,930	14,182	13,529	14,406	17,937
Florence S. M. Co..................	10,534	12,000	13,661	17,660	15,947	15,793	8,960	5,517	4,892	2,978
Shaw & Clark Sewing-Machine Co.....	2,692	3,000								
Gold Medal " " "				8,912	13,562	18,897	16,431	15,214	14,262	7,185
Davis " " "					11,568	11,376	8,861			
Domestic " "					10,397	49,554	40,114	22,700	21,452	23,587
Finkle & Lyon Mfg. Co. and Victor	2,488	2,000	1,339		2,420	7,639	7,446	6,292	6,103	5,750
Ætna Sewing-Machine Co............	2,958	3,500	4,548	5,806	4,720	4,262	3,081	1,866	1,447	707
Blees " " "					4,557	6,053	3,458			
Elliptic " " "	3,185				4,555					
Empire " " "	2,121	5,000	8,700							
Remington Sewing-Machine Co.......				3,560	2,965	4,982	9,183	17,608	25,110	12,716
Parham " " "			1,141	1,766	2,056					
Bartram & Fanton Mfg. Co.	2,958				1,004	1,000	1,000	250		
Bartlett Sewing-Machine Co.					614	1,000				
J. G. Folsom.					280					
McKay Sewing-Machine Asso.				129	218			128	161	102
C. F. Thompson.					147					
Union Buttonhole Machine Co.					124					
Leavitt Sewing-Machine Co.	1,051	1,000	771							
Goodspeed & Wyman S. M. Co.	2,126									
Keystone Sewing-Machine Co.						2,665	217	37		
Secor " " "						311	3,430	4,541	1,307	
Centennial " " "							514			

(a) Number estimated.   (b) No data.

Figure 37.—Table of sewing-machine statistics. From Frederick G. Bourne, "American Sewing Machines" in *One Hundred Years of American Commerce*, vol. 2, ed. Chauncey Mitchell Depew (New York: D.O. Haines, 1895), p. 530. (Smithsonian photo 42542–A.)

The "Sewing-Machine Combination"

WITH THE BASIC ELEMENTS of a successful sewing machine assembled, the various manufacturers should have been able to produce good machines unencumbered. The court order, however, which restrained several firms from selling Singer machines while the Howe suit was pending, started a landslide; soon Wheeler, Wilson and company, Grover and Baker company, and several others [64] purchased rights from Elias Howe. This gave Howe almost absolute control of the sewing-machine business as these companies agreed to his royalty terms of $25 for every machine sold. In an attempt to improve his own machine, Howe was almost immediately caught up in another series of legal battles in which he was the defendant; the companies he had defeated were able to accuse him of infringing on patents that they owned. To compound the confusion, individual companies also were suing each other on various grounds.

Because of this situation Orlando B. Potter, president of the Grover and Baker company, advanced in 1856 the idea of a "Combination" of sewing-machine manufacturers. He pointed out how the various companies were harming themselves by continuing litigation and tried to convince Howe that all would benefit by an agreement of some kind. He proposed that Elias Howe; Wheeler, Wilson and company; I. M. Singer and company; and Grover and Baker company pool their patents covering the essential features of the machine. The three companies had started production about the same time and approved of Potter's idea; Howe opposed it as he felt that he had the most to lose by joining the "Combination." He finally consented to take part in Potter's plan if the others would agree to certain stipulations. The first requirement was that at least twenty-four manufacturers were to be licensed. The second was that, in addition to sharing equally in the profits with the three companies, Howe would receive a royalty of $5 for each machine sold in the United States and $1 for each machine exported. It has been estimated that, as a result of this agreement, Howe received at least $2,000,000 as his share of the license fees between 1856 and 1867 when his patent expired. [65]

The organization was called the Sewing-Machine Trust and/or the Sewing-Machine Combination. The important patents contributed to it were:

1. The grooved, eye-pointed needle used with a shuttle to form the lockstitch (E. Howe patent, held by E. Howe);
2. The four-motion feeding mechanism (A. B. Wilson patent, held by Wheeler and Wilson company);
3. The needle moving vertically above a horizontal work-plate (Bachelder patent), a continuous feeding device by belt or wheel (Bachelder patent), a

[64] These included the American Magnetic Sewing Machine Co.; A. Bartholf; Nichols and Bliss; J. A. Lerow; Woolridge, Keene, and Moore; and A. B. Howe. *New York Daily Tribune,* Sept. 3, 1853.

[65] "Who Invented the Sewing-Machine," unsigned article in *The Galaxy,* vol. 4, August 31, 1867, pp. 471–481.

yielding presser resting on the cloth (Bachelder patent), the spring or curved arm to hold the cloth by a yielding pressure (Morey and Johnson patent), the heart-shaped cam as applied to moving the needle bar (Singer patent); all these patents, held by the Singer Company.[66]

The Grover and Baker company contributed several patents of relative importance, but its most important claim for admission was the fact that Potter had promoted the idea.

The consent of all four member-parties was required before any license could be granted, and all were required to have a license—even the member companies. The fee was $15 per machine. A portion of this money was set aside to pay the cost of prosecuting infringers, Howe received his initial fee, and the rest was divided between the four parties. The advantage to the licensee was that he was required to pay only one fee. Most license applications were granted; only those manufacturing a machine specifically imitating the product of a licensed manufacturer were refused.

The "Combination's" three company members each continued to manufacture, improve, and perfect

---

[66] Singer has sometimes been credited as the inventor of the various improvements covered by the patents that the Singer company purchased and later contributed to the efforts of the Combination.

its own machine. Other than the joint control of the patents, there was no pooling of interests, and each company competed to attract purchasers to buy its particular type of machine, as did the companies who were licensed by them.

In 1860, the year Howe's patent was renewed, the general license fee was reduced from $15 to $7, and Howe's special royalty was reduced to $1 per machine. Howe remained a member until his patent ran out in 1867. The other members continued the "Combination" until 1877, when the Bachelder patent, which had been extended twice, finally expired. By that time the fundamental features of the sewing machine were no longer controlled by anyone. Open competition by the smaller manufacturers was possible, and a slight reduction in price followed. Many new companies came into being—some destined to be very short-lived.

From the beginning to the end of the "Combination" there was an army of independents, including infringers and imitators, who kept up a constant complaint against it, maintaining that its existence tended to retard the improvement of the sewing machine and that the public suffered thereby. In the period immediately following the termination of the "Combination," however, only a few improvements of any importance were made, and most of these were by the member companies.

Chapter Four

Figure 38.—Gibbs' patent model, 1857.   (Smithsonian photo 45504–E.)

Less Expensive Machines

WHILE THE "COMBINATION" was attempting to solve the problems of patent litigation, another problem faced the would-be home users of this new invention. The budget limitations of the average family caused a demand for a less expensive machine, for this first consumer appliance was a most desirable commodity.[67]

There were many attempts to satisfy this demand, but one of the best and most successful grew out of a young man's curiosity. James E. A. Gibbs' first exposure to the sewing machine was in 1855 when, at the age of 24, he saw a simple woodcut illustration of a Grover and Baker machine. The woodcut represented only the upper part of the machine. Nothing in the illustration indicated that more than one thread was used, and none of the stitch-forming mechanism was visible. Gibbs assumed that the stitch was formed with one thread; he then proceeded to imagine a mechanism that would make a stitch with one thread. His solution was described in his own statement:

> As I was then living in a very out of the way place, far from railroads and public conveyances of all kinds,

modern improvements seldom reached our locality, and not being likely to have my curiosity satisfied otherwise, I set to work to see what I could learn from the woodcut, which was not accompanied by any description. I first discovered that the needle was attached to a needle arm, and consequently could not pass entirely through the material, but must retreat through the same hole by which it entered. From this I saw that I could not make a stitch similar to handwork, but must have some other mode of fastening the thread on the underside, and among other possible methods of doing this, the chainstitch occurred to me as a likely means of accomplishing the end.

> I next endeavored to discover how this stitch was or could be made, and from the woodcut I saw that the driving shaft which had the driving wheel on the outer end, passed along under the cloth plate of the machine. I knew that the mechanism which made the stitch must be connected with and actuated by this driving shaft. After studying the position and relations of the needle and shaft with each other, I conceived the idea of the revolving hook on the end of the shaft, which might take hold of the thread and manipulate it into a chainstitch. My ideas were, of course, very crude and indefinite, but it will be seen that I then had the correct conception of the invention afterwards embodied in my machine.[68]

Gibbs had no immediate interest in the sewing machine other than to satisfy his curiosity. He did not think of it again until January 1856 when he was visiting his father in Rockbridge County, Virginia. While in a tailor's shop there, he happened to see a

---

[67] *Scientific American* (Jan. 29, 1859), vol. 14, no. 21, p. 165. In a description of the new Willcox and Gibbs sewing machine the following observation is made: "It is astonishing how, in a few years, the sewing machine has made such strides in popular favor, and become, from being a mechanical wonder, a household necessity and extensive object of manufacture. While the higher priced varieties have such a large sale, it is no wonder that the cheaper ones sell in such tremendous quantities, and that our inventors are always trying to produce something new and cheap."

[68] Op. cit. (footnote 53), pp. 129–131.

Figure 39.—One of the first commercial machines produced by the Willcox & Gibbs Sewing Machine Co. in 1857, this machine bears no serial number, although the name "James E. A. Gibbs" is inscribed in two places on the cloth plate. It was used as the patent model for Gibbs' improvement on his 1857 patent issued the following year on August 10, 1858. (Smithsonian photo P. 6393.)

Singer machine. Gibbs was very much impressed, but thought the machine entirely too heavy, complicated, and cumbersome, and the price exorbitant. It was then that he recalled the machine he had devised. Remembering how simple it was, he decided to work in earnest to produce a less-expensive type of sewing machine.

Gibbs had little time to spend on this invention since his family was dependent upon him for support, but he managed to find time at night and during inclement weather. In contemporary references, Gibbs is referred to as a farmer, but since he is also reported to have had employers, it may be surmised that he was a farmhand. In any event, his decision to try to produce a less-expensive sewing machine suffered from a lack of proper tools and adequate materials. Most of the machine had to be constructed of wood, and he was forced to make his own needles. By the end of April 1856, however, his model was

sufficiently completed to arouse the interest of his employers, who agreed to furnish the money necessary to patent the machine.

Gibbs went to Washington, where he examined sewing-machine models in the Patent Office and other machines then on the market. Completing his investigations, Gibbs made a trip to Philadelphia and showed his invention to a builder of models of new inventions, James Willcox. Much impressed with the machine, Willcox arranged for Gibbs to work with his son, Charles Willcox, in a small room in the rear of his shop. After taking out two minor patents (on December 16, 1856, and January 20, 1857), Gibbs obtained his important one, U.S. patent No. 17,427 on June 2, 1857 (fig. 38). His association with Charles Willcox led to the formation of the Willcox & Gibbs Sewing Machine Company, and they began manufacturing chainstitch machines in 1857 (fig. 39). The machine used a straight needle

Figure 40.—A DOLPHIN sewing machine based on Clark's patent of 1858. This design was first used by T. J. W. Robertson in 1855, but in his patent issued on May 22 of that year no claim was made for the machine design, only for the chainstitch mechanism. The same style was used by D. W. Clark in several of his chainstitch patents, but he also made no claim for the design, stating that the machine "may be made in any desired ornamental form." The dolphin-style machines are all chainstitch models of solid brass, originally gilt. Although only about five inches long, they are full-size machines using a full-size needle. (Smithsonian photo 45505.)

to make a chainstitch. At the forward end of the main shaft was a hook which, as it rotated, carried the loop of needle-thread, elongated and held it expanded while the feed moved the cloth until the needle at the next stroke descended through the loop so held. When the needle descended through the first loop, the point of the hook was again in position to catch the second loop, at which time the first loop was cast off and the second loop drawn through it, the first loop having been drawn up against the lower edge of the cloth to form a chain.

A Gibbs sewing machine, on a simple iron-frame stand with treadle, sold for approximately $50 in the late 1850s,[69] while a Wheeler and Wilson [70] machine or a Grover and Baker [71] with the same type

of stand sold for approximately $100. After the introduction of the Gibbs machine, the Singer company [72] brought out a light family machine in 1858 that was also first sold for $100. It was then reduced to $50, but it was not popular because it was too light (see discussion of Singer machines, pp. 34–35). In 1859, Singer brought out its second, more successful family machine, which sold for $75.

Like the other companies licensed by the "Combination," Willcox and Gibbs company paid a royalty for the use of the patents it held. Although the Willcox and Gibbs machine was a single-thread chainstitch machine and the company held the Gibbs patents, the company was required to be licensed to use the basic feed, vertical needle, and other related

[69] *Scientific American*, vol. 15, no. 21 (January 29, 1859), p. 165, and Willcox and Gibbs advertising brochure, 1864.
[70] *Scientific American*, vol. 12, no. 8 (November 1, 1856), p. 62.
[71] Ibid., vol. 1, no. 19 (November 5, 1859), p. 303.

[72] I. M. Singer & Co.'s Gazette, vol. 5, no. 4 (March 1, 1859), p. 4, and a brochure, *Singer's New Family Sewing Machine* (in Singer Manufacturing Company, Historic Archives).

patents held by the "Sewing-Machine Combination."

With the approach of the Civil War, Gibbs returned to Virginia. Poor health prevented him from taking an active part in the war, but he worked throughout the conflict in a factory processing saltpeter for gunpowder. Afterward, Gibbs returned to Philadelphia and found that Willcox had faithfully protected his sewing-machine interests during his long absence. The firm prospered, and Gibbs finally retired to Virginia a wealthy man. Interestingly, Gibbs named the Virginia village to which he returned in later life "Raphine"—derived, somewhat incorrectly, from the Greek word "to sew."

The Willcox & Gibbs Sewing Machine Company is one of the few old companies still in existence. It discontinued making and selling family-style machines many years ago and directed its energies toward specialized commercial sewing machines, many of which are based on the original chainstitch principle.

There was also an ever-increasing number of other patentees and manufacturers who, in the late 1850s and 1860s, attempted to produce a sewing machine that would circumvent both the "Combination" and the high cost of manufacturing a more complicated type of machine. Some of the more interesting of these are pictured and described in figures 40 through 54.

Figure 41.—THE CHERUB sewing machine was another Robertson first which was adopted by Clark. Robertson's patent of October 20, 1857, once again makes no claim for the design; neither does Clark's patent of January 5, 1858, illustrated here. The machine is approximately the same size as the dolphin and is made in the same manner and of the same materials. Two cherubs form the main support, one also supporting the spool and leashing a dragonfly which backs the needle mechanism. (Smithsonian photo 45504–D.)

Figure 42.—THE FOLIAGE SEWING MACHINE originated
with D. W. Clark. Once again he did not include
the design in his June 8, 1858, patent, which was
aimed at improving the feeding mechanism. Like
most hand-turned models, these required a clamp
to fasten them to the table when in operation.
(Smithsonian photo 45504–C.)

Figure 44.—THE HORSE SEWING MACHINE is among
the most unusual of the patents issued for mechanical improvements. Although James Perry, the
patentee, made several claims for the looper, feeder,
and tension, he made no mention of the unusual
design of the machine, for which a patent was
issued on November 23, 1858. Although it was
probably one of a kind, the horse machine illustrates
the extent to which the inventor's mind struggled
for original design. (Smithsonian photo 45505–C.)

Figure 43.—THE SEWING SHEARS was another popular
machine of unusual style. Some models were
designed to both cut and sew, but most derived
their names from the method of motivating power.
The earliest example of the sewing-shears machine
was invented by Joseph Hendrick, who stated in
his patent that he was attempting to produce "a
simple, portable, cheap, and efficient machine."
His patent model of October 5, 1858, is illustrated.
(Smithsonian photo 45504–F.)

49

Figure 45.—Many inventors attempted to cut the cost of manufacturing a complicated machine. One of these was Albert H. Hook, whose machine is only about four inches high and two inches wide. His patent, granted November 30, 1858, simplified the construction and arrangement of the various parts. Although Hook used a barbed needle reminiscent of the one used by Thimonnier, his method of forming the stitch was entirely different. The thread was passed through the necessary guides, and when the cloth was in place the needle was thrust up from below. Passing through the fabric, the needle descended, carrying with it a loop of thread. As the process was repeated, a chainstitch was formed with the enchained loop on the under side. In spite of its simple mechanism, Hook's machine was not a commercial success. (Smithsonian photo 45505–D.)

Figure 46.—In addition to mechanical patents, a number of design patents were also issued for sewing machines. These fall into a separate series in the Patent Office's numerical records. This unusual example featured two semidraped female figures holding the spool of thread, a mermaid holding the needle, a serpent which served as the presser foot, and a heart-shaped baster plate. The design was patented by W. N. Brown, October 25, 1859, but no examples other than the patent model are known to have been made. (Smithsonian photo 45504–A.)

Figure 47.—THE SQUIRREL MACHINE was another interesting design patent. S. B. Ellithorp had received a mechanical patent for a two-thread, stationary-bobbin machine on August 26, 1857. That same month he published a picture of his machine, shown here as republished in the *Sewing Machine News*, vol. 7, no. 11, November 1885. The machine was designed in the shape of "the ordinary gray squirrel so common throughout this country— an animal that is selected as a type of provident care and forethought, for its habits of frugality and for making provision for seasons of scarcity and want in times of plenty—and the different parts of the animal are each put to a useful purpose; the moving power being placed within its body, the needle stock through its head, one of its fore feet serving to guide the thread, and the other to hold down the cloth while being sewed, and the tip of its tail forming a support to the spool from which the thread is supplied."

Although the design patent was not secured until June 7, 1859, the inventor was reported to have been perfecting his machine for manufacture in 1857. Ellithorp planned "to place them in market at a price that will permit families and individuals that have heretofore been deterred from purchasing a machine by the excessive and exorbitant price charged for those now in use, to possess one." Patent rights were sold under the name of Ellithorp & Fox, but the machine was never manufactured on a large scale, if at all. No squirrel machines are known to have survived. (Smithsonian photo 53112.)

Figure 48.—HEYER'S POCKET SEWING MACHINE patent model, November 17, 1863. This patent model is one piece, and measures about two inches in height and two inches in length. It will stitch—but only coarse, loosely woven fabrics. As can be expected, a great deal of manual dexterity is required to compensate for the omission of mechanical parts. Heyer advertised patent rights for sale, but evidence of manufactured machines of this type has yet to be discovered. (Smithsonian photo 18115–D[a].)

Figure 49.—HEYER'S MACHINE as illustrated in *Scientific American*, July 30, 1864. The smallest and most original of all the attempts to simplify machine sewing, Heyer's machine, which made a chain-stitch, was constructed of a single strip of metal. The *Scientific American* stated: "It is simply a steel spring ingeniously bent and arranged and it is said to sew small articles very well. The whole affair can easily be carried in the coat pocket."

One method of operation, vibrating with the finger, was illustrated. The machine could be operated also by holding it in the hand and pressuring it between two fingers. Cloth was inserted at *c*, and the prongs of the spring feed *f* carried it along after each stitch. It was stated that the needle could be cut from the same strip of metal, but it was advised also that the needle could be made as a separate piece and attached. (Smithsonian photo 48221.)

Figure 50.—ALTHOUGH BEAN'S AND RODGERS' running-stitch machines, the second and fourth U.S. sewing-machine patents, experienced little commercial success, small manufactured machines based on Aaron Palmer's patent of May 13, 1862, were popular in the 1860s. The patent model above is a small brass implement with crimping gears that forced the fabric onto an ordinary sewing needle. The full needle was then removed from its position, and the thread was pulled through the fabric by hand. (Smithsonian photo 45524.)

Figure 51.—ONE OF THE EARLY COMMERCIAL MANUFACTURERS of the Palmer patent was Madame Demorest, a New York dressmaker. She advertised her Fairy sewing machine in *Godey's Lady's Book*, vol. 66, 1863, and stated: "In the first place it will attract attention from its diminutive, fairy-like size, and with the same ease with which it can be carried, an important matter to a seamstress or dressmaker employed from house to house . . . . What no other sewing machine attempts to do, it runs, and does not stitch, it sews the more delicate materials an ordinary sewing machine cuts or draws . . . ." (Smithsonian photo 43690.)

Figure 52.—THE FAIRY SEWING MACHINE sold for five dollars and was adequate for its advertised purpose, sewing or running very lightweight fabrics. The machine was marked with the Palmer patent, the date May 13, 1862, and the name "Mme. Demorest."

A machine identical to the Fairy, but bearing both Palmer patent dates, May 13, 1862, and June 19, 1863, and the name "Gold Medal," was manufactured by a less-scrupulous company. This machine was advertised as follows: "A first class sewing machine, handsomely ornamented, with all working parts silver plated. Put up in a highly polished mahogany case, packed ready for shipment. Price $10.00. This machine uses a common sewing needle, is very simple. A child can operate it. Cash with order." Some buyers felt they were swindled, as they had expected a heavy-duty machine, but no recourse could be taken against the advertiser. Another similar machine was also manufactured under the name "Little Gem." (Smithsonian photo 45525.)

Figures 53 and 54.—RUNNING-STITCH MACHINES were also attempted by several other inventors. Shaw & Clark, manufacturers of chainstitch machines, patented this running-stitch machine on April 21, 1863. From the appearance of the patent model, it was already in commercial production. On May 26, 1863, John D. Dale also received a patent for an improvement related to the method of holding the needle and regulating the stitches in a running-stitch machine. Dale's patent model was a commercial machine.

John Heberling patented several improvements in 1878 and 1880. His machine, which was a little larger and in appearance resembled a more conventional type of sewing machine, was a commercial success. (Shaw & Clark: Smithsonian photo P. 6395; Dale: Smithsonian photo P. 6394.)

DALE

Appendixes

I. Notes on the Development and Commercial Use of the Sewing Machine

INTRODUCTION

While researching the history of the invention and the development of the sewing machine, many items of related interest concerning the machine's economic value came to light. The manufacture of the machines was in itself a boost to the economy of the emerging "industrial United States," as was the production of attachments for specialized stitching and the need for new types of needles and thread. Moreover, the machine's ability to speed up production permitted it to permeate the entire field of products manufactured by any type of stitching, from umbrellas to tents. Since this aspect of the story was not completed for this study, no attempt will be made to include any definitive statements on the economic importance of the sewing machine at home or abroad. This related information is of sufficient interest, however, to warrant inclusion in this first Appendix. Perhaps these notes will suggest areas of future research for students of American technology.

READY—MADE CLOTHING

Whether of the expensive or the inexpensive type, the sewing machine was much more than a popular household appliance. Its introduction had far-reaching effects on many different types of manufacturing establishments as well as on the export trade. The newly developing ready-made clothing industry was not only in a state of development to welcome the new machine but also was, in all probability, responsible for its immediate practical application and success.

Until the early part of the second quarter of the 19th century, the ready-made clothing trade in the United States was confined almost entirely to furnishing the clothing required by sailors about to ship out to sea. The stores that kept these supplies were usually in the neighborhood of wharf areas. But other than the needs of these seamen, there was little market for ready-made goods. Out of necessity many of the families in the early years in this country had made their own clothing. As wealth was acquired and taste could be cultivated, professional seamstresses and tailors were in increasing demand, moved into the cities and towns, and even visited the smaller villages for as long as their services were needed. At the same time a related trade was also growing in the cities, especially in New York City, that of dealing in second-hand clothing. Industrious persons bought up old clothes, cleaned, repaired and refinished them, and sold the clothing to immigrants and transients who wished to avoid the high cost of new custom-made clothing.

The repairing of this second-hand clothing led to the purchase of cheap cloth at auction—"half-burnt," "wet-goods," and other damaged yardage. When in excess of the repairing needs, this fabric was made into garments and sold with the second-hand items. Many visitors who passed through New York City were found to be potential buyers of this merchandise if a better class of ready-made clothes was made available. Manufacture began to increase. Tailors of the city began to keep an assortment of finished garments on hand. When visitors bought these, they were also very likely to buy additional garments for resale at home. The latter led to the establishment of the wholesale garment-manufacturing industry in New York about 1834–35.

Most of the ready-made clothing establishments were small operations, not large factories. Large quantities of cloth were purchased; cutting was done in multiple layers with tailor's shears. Since many seamstresses were needed, the garments were farmed out to the girls in their homes. The manufacture of garments in quantity meant that the profit on each garment was larger than a tailor could make on a single custom-made item. The appeal of increased profits influenced many to enter the new industry and, due to the ensuing competition, the retail cost of each garment was lowered. Just as the new businesses were getting underway, the Panic of 1837 ruined most of them. But the lower cost and the convenience of ready-made clothing had left its mark. Not only was the garment-manufacturing business re-established soon after the Panic had subsided, but by 1841 the value of clothing sold at wholesale in New York was estimated at $2,500,000 and by 1850—a year before sewing machines were manufactured in any quantity—there were 4,278 clothing manufacturing establishments in the United States. Beside New York City, Cincinnati was also one of the important ready-made clothing centers. In 1850 the value of its products amounted to $4,427,500 and in 1860 to $6,381,190. Boston was another important center with a ready-made clothing production of $4,567,749 in 1860. Philadelphia, Baltimore, Louisville, and St. Louis all had a large wholesale clothing trade by 1860. Here was the ready market for a practical sewing machine.[73]

Clothing establishments grew and began to have agencies in small towns and the sewing work was distributed throughout the countryside. The new, competing sewing-machine companies were willing to deliver a machine for a small sum and to allow the buyer to pay a dollar or two a month until the full amount of the sale was paid. This was an extension of the hire-purchase plan (buying on credit) initiated by Clark of the Singer Company. The home seamstresses were eager to buy, for they were able to produce more piecework with a sewing machine and therefore earn more money. An example of the effect that the sewing machine had on the stitching time required was interestingly established through a series of experiments conducted by the Wheeler and Wilson company. Four hand sewers and four sewing-machine operators were used to provide the

average figures in this comparative time study, the results of which were published in 1861; [74]

NUMBER OF STITCHES PER MINUTE

	By Hand	By Machine
Patent leather, fine stitching...........	7	175
Binding hats..........	33	374
Stitching vamped shoes.	10	210
Stitching fine linen.....	23	640
Stitching fine silk......	30	550

TIME FOR GARMENTS STITCHED

	By Hand	By Machine
Frock coats......	16 hrs. 35 min.	2 hrs. 38 min.
Satin vests.......	7 hrs. 19 min.	1 hr. 14 min.
Summer pants...	2 hrs. 50 min.	0 hr. 38 min.
Calico dress......	6 hrs. 37 min.	0 hr. 57 min.
Plain apron......	1 hr. 26 min.	0 hr. 9 min.
Gentlemen's shirts..........	14 hrs. 26 min.	1 hr. 16 min.

The factory manufacturer, with the sewing work done at the factory, was also developing. In 1860, Oliver F. Winchester, a shirt manufacturer of New Haven, Connecticut, stated that his factory turned out 800 dozen shirts per week, using 400 sewing machines and operators to do the work of 2,000 hand sewers. The price for hand sewing was then $3 per week, which made labor costs $6000 per week. The 400 machine operators received $4 per week, making the labor cost $1600 per week. Allowing $150 as the cost of each machine, the sewing machines more than paid for themselves in less than 14 weeks, increased the operators pay by $1 a week, and lowered the retail cost of the item.[75] The greatest savings of time, which was as much as fifty percent, was in the manufacture of light goods—such items as shirts, aprons, and calico dresses. The Commissioner of Patents weighed the monetary effect that this or any invention had on the economy against the monetary gain received by the patentee. When he found that the patentee had not been fairly compensated, he had the authority to grant a seven-year extension to the patent.[76]

[73] *Eighth Census, 1860, Manufactures, Clothing* (United States Census Office, published Government Printing Office: Washington, D.C., 1865).

[74] *Eighty Years of Progress of the United States* (New York, 1861), vol. 2, pp. 413–429.

[75] GEORGE GIFFORD, "Argument of [George] Gifford in Favor of the Howe Application for Extension of Patent" (New York: United States Patent Office, 1860).

[76] Op. cit. (footnote 34).

The sewing machine also contributed to the popularity of certain fashions. Ready-made cloaks for women were a business of a few years' standing when the sewing machine was adopted for their manufacture in 1853. Machine sewing reduced the cost of constructing the garment by about eighty percent, thereby decreasing its price and increasing its popularity. In New York City alone, the value of the "cloak and mantilla" manufacture in 1860 was $618,400.[77] Crinolines and hoopskirts were easier to stitch by machine than by hand, and these items had a spirited period of popularity due to the introduction of the sewing machine. Braiding, pleating, and tucking adorned many costume items because they could be produced by machine with ease and rapidity.

In addition to using the sewing machine for the manufacture of shirts, collars, and related men's furnishings, the machine was also used in the production of men's and boy's suits and reportedly gave "a vast impetus to the trade."[78] The Army, however, was not quite convinced of the sewing machine's practical adaptation to its needs. Although a sewing machine was purchased for the Philadelphia Quartermaster Depot as early as 1851, they had only six by 1860. On March 31, 1859, General Jesup of the Philadelphia Depot wrote to a Nechard & Company stating that the machine sewing had been tried but was not used for clothing, only for stitching caps and chevrons. In another letter, on the same day, to "Messers Hebrard & Co., Louisiana Steam Clothing Factory, N. Orleans," Jesup states: "Machine sewing has been tried with us, and though it meets the requirements of a populous and civilized life, it has been found not to answer for the hard wear and tear and limited means of our frontier service. Particular attention has been paid to this subject, and we have abandoned the use of machines for coats, jackets and trousers, etc and use them on caps and bands that are not exposed to much hard usage. . . ."[79] At this period prior to the Civil War, the Army manufactured its own clothing. As the demands of war increased, more and more of the Army's clothing supplies were furnished on open contract—with no

Figure 55.—Blake's leather-stitching machine patent model of July 6, 1858; the inventor claimed the arrangement of the mechanism used and an auxiliary arm capable of entering the shoe, which enabled the outer sole to be stitched both to the inner sole and to the upper part of the shoe. (Smithsonian photo 50361.)

specifications as to stitching.[80] Machine stitching, in fact, is found in most of the Civil War uniforms. One of the problems that most probably affected the durability of the machine stitching in the 1850s was the sewing thread, a problem that was not solved until the 1860s and which is discussed later under "thread for the machine."

SHOE MANUFACTURE

Another industry that was aided by the new invention was that of shoe manufacture. Although the earliest sewing-machine patents in the United States reflect the inventors' efforts to solve the difficult task of leather stitching, and, although machines were used to a limited extent in stitching some parts of

[77] Eighth Census, 1860, Manufactures (United States Census Office, published Government Printing Office: Washington, D.C., 1865), "Women's Ready-Made Clothing," p. 83.

[78] Ibid., p. 64.

[79] National Archives, Record Group 92, Office of the Quartermaster General, Clothing Book, Letters Sent, volume 17.

[80] The author wishes to acknowledge the valuable help of Mr. Donald Kloster of the Smithsonian Institution's Division of Military History for the preceding four references and related information.

Figure 56.—HARRIS' patent thread cutter, 1872.
(Smithsonian photo P-6397.)

Figure 57.—WEST's patent thread cutter, 1874.
(Smithsonian photo P-63100.)

Figure 58.—KARR's patent needle threader, 1871.
(Smithsonian photo P-63101.)

the shoe in the early and mid-1850s, it was not until 1858 that a machine was invented that could stitch the sole to the inner sole and to the upper part of the shoe. This was the invention of Lyman R. Blake and was patented by him on July 8, 1858; the patent model is shown in figure 55. Blake formed a chain-stitch by using a hooked needle, which descended from above, to draw a thread through the supporting arm. Serving as the machine's bedplate, the arm was shaped to accommodate the stitching of all the parts of the shoe.

The increased number of shoes required by the Army during the Civil War spurred the use of the sewing machine in their manufacture. The first "machine sewed bootees" were purchased by the Army in 1861. Inventors continued their efforts; the most prominent of these was Gordon McKay, who worked on an improvement of the Blake machine with Robert Mathies in 1862 and then with Blake in 1864. Reportedly, the Government at first preferred the machine-stitched shoes as they lasted eight times longer than those stitched by hand; during the war

the Army purchased 473,000 pairs, but in 1871 the Quartermaster General wrote:

> No complaints regarding the quality of these shoes were received up to February 1867 when a Board of Survey, which convened at Hart's Island, New York Harbor reported upon the inferior quality of certain machine sewed bootees of the McKay patent, issued to the enlisted men at that post. The acting Quartermaster General, Col. D. H. Rucker, April 10, 1867, addressed a letter to all the officers in charge of depots, with instructions not to issue any more of the shoes in question, but to report to this office the quantity remaining in store. From these reports it appears that there were in store at that time 362,012 pairs M. S. Bootees, all of which were ordered to be, and have since been sold at public auction.[81]

The exact complaint against the shoes was not recorded. Possibly the entire shoe was stitched by machine. It was found that although machine-stitched shoes were more durable in some respects and the upper parts of most shoes continued to be machine stitched, pegged soles for the more durable varieties remained the fashion for a decade or more, as did custom hand-stitched shoes for those who could afford them.

OTHER USES

The use of sewing machines in all types of manufacturing that required stitching of any type continued to grow each year. While the principal purpose for which they were utilized continued to be the manufacture of clothing items, by the year 1900 they were also used for awnings, tents, and sails; cloth bags; bookbinding and related book manufacture; flags and banners; pocketbooks, trunks, and valises; saddlery and harnesses; mattresses; umbrellas; linen and rubber belting and hose; to the aggregate sum of nearly a billion dollars—$979,988,413.[82]

SEWING-MACHINE ATTACHMENTS

The growing popularity of the sewing machine offered still another boost to the economy—the development of many minor, related manufacturing

---

[81] Letter of Nov. 4, 1871, to Col. Theo. A. Dodge, USA (Ret.), Boston, from Quartermaster General M. C. Meigs, in the National Archives, Record Group 92, Quartermaster General's Office, Letters Sent, Clothing Supplies, 1871.

[82] *Twelfth Census of the United States, 1900*, vol. 10, *Manufactures*, Part 4, Special Reports on Selected Industries (United States Census Office, Washington, D.C., 1902).

Figure 59.—SHANK's patent bobbin winder, 1870.
(Smithsonian photo P-6398.)

Figure 60.—SWEET's patent binder, 1853.
(Smithsonian photo P-6396.)

Figure 61.—SPOUL's patent braid guide, 1871.
(Smithsonian photo P-63102.)

Figure 62.—ROSE's patent embroiderer, 1881.
(Smithsonian photo P-6399.)

Figure 63.—HARRIS' patent buttonhole attachment,
1882. (Smithsonian photo P-63103.)

industries. The repetitive need for machine needles, the development of various types of attachments to simplify the many sewing tasks, and the ever-increasing need for more and better sewing thread—the sewing machine consumed from two to five times as much thread as stitching by hand—created new manufacturing establishments and new jobs.

The method of manufacturing machine needles did not differ appreciably from the method used in making the common sewing needle, but the latter had never become an important permanent industry in the United States. Since the manufacture of practical sewing machines was essentially an American development and the eye-pointed needle a vital component of the machine, it followed that the manufacture of needles would also develop here. Although such a manufacture was established in 1852,[83] foreign imports still supplied much of the

[83] CHARLES M. KARCH, "Needles: Historical and Descriptive," in *Twelfth Census of the United States, 1900*, vol. 10, *Manufactures*, Part 4, Special Reports on Selected Industries (United States Census Office: Washington, D.C., 1902), pp. 429–432.

Figure 64.—THE TREADLE OF THE MACHINE was also used to help create music. George D. Garvie and George Wood received patent 267,874, Nov. 21, 1882, for "a cover for a sewing machine provided with a musical instrument and means for transmitting motion from the shaft of the sewing machine to the operating parts of the musical instrument." Although no patent model was submitted by the inventors, the "Musical Sewing Machine Cover" was offered for sale as early October 1882, as shown by this advertisement that appeared in *The Sewing Machine News* that month. (Smithsonian photo 57983.)

Figure 65.—THIS FANNING ATTACHMENT was commercially available from James Morrison & Co. in the early 1870s; it sold for one dollar as stated in the advertising brochure from which this engraving was copied. Other inventors also patented similar implements. (Smithsonian photo 45513.)

need in the 1870s. As more highly specialized stitching machines were developed, an ever-increasing variety of needles was required, and the industry grew.

Soon after the sewing machine was commercially successful, special attachments for it were invented and manufactured. These ranged from the simplest devices for cutting thread to complicated ones for making buttonholes (see figs. 56 through 66).

The first patent for an attachment was issued in 1853 to Harry Sweet for a binder, used to stitch a special binding edge to the fabric. Other related attachments followed; among these were the hemmer which was similar to the binder, but turned the edge of the same piece of fabric to itself as the stitching was performed. Guides for stitching braid in any pattern, as directed by the movement of the goods below, were also developed; this was followed by the embroiderer, an elaborate form of braider. The first machine to stitch buttonholes was patented in 1854 and the first buttonhole attachment in 1856, but the latter was not practical until improvements were made in the late 1860s. Special devices for refilling the bobbins were invented and patented as early as 1862, and the popularity of tucked and ruffled garments inspired inventors to provide sewing-machine attachments for these purposes also. To keep the seamstress cool, C. D. Stewart patented an attachment for fanning the operator by an action derived from the treadle (fig. 65). While electric

Figure 66.—JONES "ELECTRO MOTOR" PATENT MODEL of 1871 on a Bartlett sewing machine. (Smithsonian photo P-63104.)

Figure 67.—SIX-CORD cabled thread.

sewing machines did not become common until the 20th century, several 19th-century inventors considered the possibility of attaching a type of motor to the machine. One was the 1871 patent of Solomon Jones, who added an "electro motor" to an 1865 Bartlett machine (fig. 66). The attachments that were developed during the latter part of the 19th century numbered in the thousands; many of these were superfluous. Most of the basic ones in use today were developed by the 1880s and remain almost unchanged. Even the recently popular home zigzag machine, an outgrowth of the buttonhole machine, was in commercial use by the 1870s.

Sewing-machine improvements have been made from time to time. Like other mechanical items the machine has become increasingly automatic, but the basic principles remain the same. One of the more recent developments, patented [84] in 1933 by Valentine Naftali et al., is for a manufacturing machine that imitates hand stitching. This machine uses a two-pointed "floating needle" that is passed completely through the fabric—the very idea that was attempted

over one hundred years ago. The machine is currently used by commercial manufacturers to produce decorative edge-stitching that very closely resembles hand stitching.

THREAD FOR THE MACHINE

The need for a good thread durable enough to withstand the action of machine stitching first created a problem and ultimately another new industry in this country. When the sewing machine was first developed the inventors necessarily had to use the sewing thread that was available. But, although the contemporary thread was quite suitable for hand sewing, it did not lend itself to the requirements of the machine. Cotton thread, then more commonly a three-ply variety, had a glazed finish and was wiry. Silk thread frequently broke owing to abrasion at the needle eye. For the most part linen thread was too coarse, or the fine variety was too expensive. All of the thread had imperfections that went unnoticed in the hands of a seamstress, but caused havoc in a machine.

[84] U.S. patent 1,931,447, issued to Valentine Naftali, Henry Naftali, and Rudolph Naftali, Oct. 17, 1933. The Naftali machines are manufactured by the American Machine and Foundry Company and are called AMF Stitching Machines.

Quality silk thread that would withstand the rigors of machine stitching could be produced, but it was quite expensive also. A new type of inexpensive thread was needed; the obvious answer lay in improving the cotton thread.[85]

In addition to the popular three-ply variety, cotton thread was also made by twisting together either two single yarns or more than three yarns. Increasing the number of yarns produced a more cylindrical thread. The earliest record of a six-ply cotton thread was about 1840.[86] And in 1850 C. E. Bennett of Portsmouth, New Hampshire, received a gold medal for superior six-cord, or six-ply, spool cotton at the Fair of the American Institute. But the thread was still wiry and far from satisfactory. By the mid-1860s the demonstrated need for thread manufacturers in America brought George A. Clark and William Clark, third generation cotton-thread manufacturers of Paisley, Scotland, to Newark, New Jersey, where they built a large mill. George Clark decided that a thread having both a softer finish and a different construction was needed. He produced a six-cord cabled thread, made up of three two-ply yarns (see fig. 67). The thread was called "Clark's 'Our New Thread,'" which was later shortened to O.N.T. The basic machine-thread problem was solved. When other manufacturers used the six-cord cabled construction they referred to their thread as "Best Six-Cord"[87] or "Superior Six-Cord"[88] to distinguish it from the earlier variety made up of six single yarns in a simple twist. Another new side industry of the sewing machine was successfully established.

MANUFACTURE AND EXPORT, TO 1900

Sewing machines were a commodity in themselves, both at home and abroad. In 1850, there were no establishments exclusively devoted to the manufacture of sewing machines, the few constructed were made in small machine shops. The industry, however, ex-perienced a very rapid growth during the next ten years. By 1860 there were 74 factories in 12 States,[89] mainly in the East and Midwest,[90] producing over 111,000 sewing machines a year. In addition, there were 14 factories that produced sewing-machine cases and attachments. The yearly value of these products was approximately four and a half million dollars, of which the amount exported in 1861 was $61,000. Although the number of sewing-machine factories dropped from 74 in 1860 to 69 by 1870, the value of the machines produced increased to almost sixteen million dollars.

The number of sewing-machine companies fluctuated greatly from year to year as many attempted to enter this new field of manufacture. Some were not able to make a commercial success of their products. The Civil War did not seem to be an important factor in the number of companies in business in the North. Although one manufacturer ceased operations in Richmond, Virginia, and a Vermont firm converted to arms manufacture, several companies began operations during the war years. Of the 69 firms in business in 1870, only part had been in business since 1860 or before; some were quite new as a result of the expiration of the Howe patent renewal in 1867.

Probably due to the termination of many of the major patents, there were 124 factories in 1880, but the yearly product value remained at sixteen million dollars. The 1890 census reports only 66 factories with a yearly production of a little less than the earlier decade. But by 1900, the yearly production of a like number of factories had reached a value of over twenty-one million, of which four and a half million dollars worth were exported annually. The total value of American sewing machines exported from 1860 to 1900 was approximately ninety million dollars. The manufacture of sewing machines made a significant contribution to the economic development of 19th-century America.

---

[85] See Appendix V, p. 135, "A Brief History of Cotton Thread."

[86] *The Story of Cotton Thread* (New York, The Spool Cotton Company, 1933).

[87] J. and P. Coats spool cotton.

[88] Willimantic spool cotton.

---

[89] New Hampshire, Vermont, Massachusetts, Rhode Island, Connecticut, New York, Pennsylvania, Delaware, Ohio, Indiana, Illinois, Kentucky. *Eighth Census, 1860, Manufactures* (United States Census Office, published by Government Printing Office: Washington, D.C., 1865.)

[90] Sewing-machine manufacture in the South was just beginning to blossom when it was curtailed by the outbreak of the Civil War. See Lester sewing machine, figure 109 on page 102.

II. American Sewing-Machine Companies
of the 19th Century

During the latter half of the 19th century, there was a total of two hundred or more sewing-machine companies in the United States. Although a great many manufacturing-type machines were sold, this business was carried on by relatively few companies and most were primarily concerned with the family-type machines. A representative number of these family machines together with information concerning both the company and serial-number dating are found in figures 68 through 132. A great many of the companies were licensed by the "Combination," but, in addition, some companies were constructing machines that did not infringe the patents, other companies infringed the patents but managed to avoid legal action, and there were numerous companies that mushroomed into existence after the "Combination" was dissolved in 1877. Most of the latter were very short-lived. It is difficult to establish the exact dates of some of these companies as many of their records were incomplete or have since disappeared; even a great many of the "Combination" records were lost by fire. A summary of the existing records kept by the "Combination" is given in figure 37.

As will be noted in the subsequent listing, only a small percentage of the companies were in business for a period longer than ten years; of those that continued longer, all but a few had disappeared by 1910. Today there are about sixty United States sewing-machine companies. Most of them manufacture highly specialized sewing machines used for specific types of commercial work; only a few produce family or home-style machines. Foreign competition has increased, and the high cost of skilled labor in this country has made competition in this consumer-product field increasingly difficult. The countless varieties of American family sewing machines, so evident in the 19th century, have passed away.

Sewing Machine	Manufacturer or Company	First Made or Earliest Record	Discontinued or Last Record
Aetna	Aetna Sewing Machine Co., Lowell, Mass.	ca.1867	ca.1877
Aiken and Felthousen	——, Ithaca, N.Y.	ca.1855	before 1880
Alsop	——	—	ca.1880
American	American Sewing Machine Co.	1854	—
American Buttonhole, Overseaming and Sewing Machine (fig. 68)	American Buttonhole, Overseaming and Sewing Machine Co., Philadelphia, Pa.	1869	ca.1874
Later New American (fig. 69)	American Sewing Machine Co., Philadelphia, Pa.	ca.1874	ca.1886
American Magnetic (fig. 70)	American Magnetic Sewing Machine Company, Ithaca, N.Y.	1853	1854

Sewing Machine	Manufacturer or Company	First Made or Earliest Record	Discontinued or Last Record
Atlantic (fig. 71)	——	1869	ca.1870
Atwater (fig. 87)	——	1857	ca.1860
Avery	Avery Sewing Machine Co., New York, N.Y.	1852	185–
Avery	Avery Manufacturing Co., New York, N.Y.	1875	1886–1900
A. Bartholf Manfr. Blodgett & Lerow patent 1849 (see also)	A. Bartholf, manufacturer, New York, N.Y.	ca.1850	185–
A. Bartholf Manfr. Howe's patent, 1846 (fig. 72)	A. Bartholf, manufacturer, New York, N.Y.	1853	ca.1856
Bartholf	A. Bartholf, manufacturer Bartholf Sewing Machine Co.	1857 1859	1859 ca.1865
Bartlett (fig. 73)	Goodspeed & Wyman Bartlett Sewing Machine Co., New York, N.Y.	1866 ca.1870	ca.1870 1872
Baker	——	—	before 1880
Bartram & Fanton (fig. 74)	Bartram & Fanton Mfg. Co., Danbury, Conn.	1867	1874
Bay State	——	—	before 1880
Beckwith (fig. 75)	Barlow & Son, New York, N.Y. Beckwith Sewing Machine Co., New York, N.Y.	1871 1872	1872 ca.1876
Blees	Blees Sewing Machine Co.	1870	1873
Blodgett & Lerow (fig. 21)	O. Phelps, Boston, Mass. Goddard, Rice & Co., Worcester, Mass.	1849 1849	1849 1850
(fig. 20)	A. Bartholf, manufacturer, New York, N.Y.	1849	185–
Bond	——	—	before 1880
Boston Later New Boston	J. F. Paul & Co., Boston, Mass. Boston Sewing Machine Co., Boston, Mass.	1880 —	— after 1886
Boudoir (fig. 76)	Daniel Harris, inventor and patentee Manufacturer—several	1857	ca.1870
Bradford & Barber	Bradford & Barber, manufacturers, Boston, Mass.	1860	1861
Brattleboro	Samuel Barker and Thomas White, Brattleboro, Vt.	ca.1858	1861
Buckeye Later New Buckeye (fig. 77) (see Wilson)	Wilson [W. G.] Sewing Machine Company, Cleveland, Ohio	ca.1867	ca.1876

Sewing Machine	Manufacturer or Company	First Made or Earliest Record	Discontinued or Last Record
Buell, "E. T. Lathbury's Patent"	A. B. Buell, Westmoreland, New York	ca.1860	—
Burnet & Broderick	Burnet, Broderick and Co.	1859	ca.1860
Centennial (fig. 78)	Centennial Sewing Machine Co. (see McLean and Hooper), Philadelphia, Pa.	1873	1876
Chamberlain	Woolridge, Keene and Moore, Lynn, Mass.	1853	ca.1854
Chicago Singer	Scates, Tryber & Sweetland Mfg. Co., Chicago, Ill.	1879	1882
Later Chicago	Chicago Sewing Machine Co.	1882	ca.1885
Chicopee (see Shaw & Clark)			
Clark (fig. 42)	D. W. Clark, Bridgeport, Conn.	ca.1858	after 1860
Clark's Revolving Looper [double thread] (fig. 79) (see Windsor)	Lamson, Goodnow & Yale, Windsor, Vt.	1859	1861
Clinton	Clinton Brothers, Ithaca, N.Y.	ca.1861	ca.1865
Companion	Thurston Mfg. Co., Marlboro, N.H.	1882	—
Crown (see Florence)	Florence Sewing Machine Co., Florence, Mass.	1879	after 1886
Dauntless (later New Dauntless)	Dauntless Mfg. Co., Norwalk, Ohio	1877	after 1882
Davis	J.A. Davis, New York, N.Y.	ca.1860	—
Davis Vertical Feed	Davis Sewing Machine Co., Watertown, N.Y.	1869	after 1886
Davis Vertical Feed and Rotary Shuttle	Davis Sewing Machine Co., Dayton, Ohio	after 1886	1924
Decker (also The Princess)	Decker Mfg. Co., Detroit, Mich.	—	before 1881
Demorest	Demorest Mfg. Co. (formerly N.Y. Sewing Machine Co.)	1882	1908
Diamond (formerly Sigwalt)	Sigwalt Sewing Machine Co., Chicago, Ill.	1880	—
Domestic	Wm. A. Mack & Co. and N. S. Perkins, Norwalk, Ohio	1864	1869
Domestic	Domestic Sewing Machine Co., Norwalk, Ohio, acquired by White Sewing Machine Co. in 1924 and maintained as a subsidiary at Cleveland, Ohio.	1869	*

*Still in existence.

Sewing Machine	Manufacturer or Company	First Made or Earliest Record	Discontinued or Last Record
Dorcas	John P. Bowker, Boston, Mass.	1853	185–
Du Laney (fig. 80) Also called Little Monitor (see)			
Durgin	Charles A. Durgin, New York, N.Y.	1853	after 1855
Eldredge	Eldredge Sewing Machine Co., Chicago, Ill.	1869	1890
Elliptic			
Sloat's Elliptic	George B. Sloat and Co., Philadelphia, Pa.	ca.1858	ca.1860
Sloat's Elliptic	Union Sewing Machine Co., Richmond, Va.	1860	1861
Elliptic	Wheeler & Wilson Mfg. Co.	1861	ca.1867
	Elliptic Sewing Machine Co., N.Y., N.Y.	1867	before 1880
Empire (fig. 86) Later Remington-Empire	Empire Sewing Machine Co., Boston, Mass.	ca.1860	1869
Empress	Manufactured on order through Jerome B. Secor, Bridgeport, Conn.	1877	—
Estey	Estey Sewing Machine Co.	ca.1880	1882
Estey, Fuller-Model	Brattleboro Sewing Machine Co., Brattleboro, Vt.	1883	after 1886
Eureka (fig. 81)	Eureka Shuttle Sewing Machine Co., New York, N.Y.	1859	—
Excelsior	Excelsior Sewing Machine Co., New York, N.Y.	1854	1854
Fairy (figs. 51, 52)	Madame Demorest, New York, N.Y.	1863	ca.1865
Finkle, M. (fig. 82)	M. Finkle, Boston, Mass.	1856	ca. 1859
Finkle & Lyon	Finkle & Lyon Sewing Machine Co., Boston, Mass.	ca.1859	1867
Later Victor			
First and Frost	First and Frost, New York, N.Y.	ca.1859	ca.1861
Florence (fig. 83) Later Crown	Florence Sewing Machine Co., Florence, Mass.	ca.1860	after 1878
Folsom (see Globe and New England)	Folsom, J. G., Winchendon, Mass.	1865	ca.1871
Fosket and Savage	Fosket and Savage, Meriden, Conn.	1858	1859
Foxboro	Foxboro Rotary Shuttle Co., Foxboro, Mass.	ca.1882	—
Franklin	Franklin Sewing Machine Co., Mason Village, N.H.	1871	1871

Sewing Machine	Manufacturer or Company	First Made or Earliest Record	Discontinued or Last Record
Free	Free Sewing Machine Co., Chicago and Rockford, Ill.	1898	*
Gardner	C. R. Gardner, Detroit, Mich.	1856	—
Globe (figs. 84, 85)	J. G. Folsom, Winchendon, Mass.	1865	1869
Gold Medal (chainstitch)	Gold Medal Sewing Machine Co., Orange, Mass.	1863	1876
Gold Medal (running stitch)	——	1863	ca.1865
Gold Hibbard	Hibbard, B. S., & Co.	1875	—
Goodbody (sewing shears)	Goodbody Sewing Machine Co., Bridgeport, Conn.	1880	ca.1890
Goodes	Rex & Bockius, Philadelphia, Pa.	ca.1876	before 1881
Goodrich	H. B. Goodrich, Chicago, Ill.	ca.1880	ca.1895
Grant Brothers (fig. 90)	Grant Bros. & Co., Philadelphia, Pa.	1867	ca.1870
Greenman and True (fig. 91)	Greenman and True Mfg. Co. Norwich, Conn.	1859	1860
	Morse and True	1860	1861
Green Mountain	——	ca.1860	—
Griswold Variety	L. Griswold, New York, N.Y.	ca.1886	ca.1890
Grover and Baker (figs. 34–36, 92)	Grover and Baker Sewing Machine Co., Boston, Mass.	1851	1875
Hancock (figs. 93, 94)	——	1868	before 1881
Heberling Running Stitch	John Heberling	1878	ca.1885
Herron's Patent (fig. 95)	——	1857	—
Higby Later Acme	Higby Sewing Machine Co., Brattleboro, Vt.	ca.1882	after 1886
Home Home Shuttle	Johnson, Clark & Co., Orange, Mass.	1869	after 1876
Homestead	——	ca.1881	—
Household	Providence Tool Co., Providence, R.I.	1880	ca.1884
	Household Sewing Machine Co.	ca.1885	1906

Howe (figs. 96, 97)	Howe Sewing Machine Co., New York, N.Y.
	    (company of A. B. Howe sold
	      to Howe Machine Co.)
		1853	1873

Howe (fig. 98)	Howe Machine Co., Bridgeport, Conn.	1867	1886
Howe's Improved Patent (fig. 107)	Nichols and Bliss, Boston, Mass.	1852	1853
	J. B. Nichols & Co.	1853	1854
which became Leavitt	Nichols, Leavitt & Co., Boston, Mass.	1854	1856

*Still in existence.

Sewing Machine	Manufacturer or Company	First Made or Earliest Record	Discontinued or Last Record
N. Hunt, which became Hunt and Webster (figs. 99, 100)	N. Hunt & Co., Boston, Mass.	1853	1854
	Hunt and Webster, Boston, Mass.	1854	1857
Later Ladd and Webster (*see*)			
Improved Common Sense (fig. 102)	——	ca.1870	—
Independent Noiseless	Independent Sewing Machine Co., Binghamton, N.Y.	1873	—
Jennie June	June Mfg. Co., Chicago, Ill. Later Belvidere, Ill.	1881	1890
Jewel	Jewel Mfg. Co., Toledo, Ohio	1884	after 1886
Johnson (fig. 103)	Emery, Houghton & Co., Boston, Mass.	1856	after 1865
Keystone	Keystone Sewing Machine Co.	before 1872	ca.1874
Ladd & Webster (fig. 101)	Ladd, Webster & Co., Boston, Mass.	1858	ca.1866
Ladies Companion (fig. 115) (*see* Pratt's Patent)	——	1858	ca.1858
"Lady" (fig. 104)	——	1859	—
Landfear's Patent (fig. 105)	Parkers, Snow, Brooks & Co., West Meriden, Conn.	1857	—
Langdon	L. W. Langdon	1856	—
Lathrop (fig. 106)	Lathrop Combination Sewing Machine Co.	1873	—
Leader	Leader Sewing Machine Co., Springfield, Mass.	1882	—
Leavitt (fig. 108)	Nichols, Leavitt & Co., Boston, Mass.	1855	1857
	Leavitt & Co.	1857	ca.1865
	Leavitt Sewing Machine Co.	ca.1865	1870
Leslie Revolving Shuttle	Leslie Sewing Machine Co., Cleveland, Ohio	1881	—
Lester (fig. 109)	J. H. Lester, Brookyln, N.Y.	ca. 1858	early 1860
	Lester Mfg. Co., Richmond, Va.	early 1860	late 1860
	Union Sewing Machine Co., Richmond, Va.	late 1860	1861
Little Gem	——	—	ca.1870
Little Giant	Domestic Sewing Machine Co., Norwalk, Ohio	ca.1882	—
Little Monitor (not associated with Monitor)	G. L. Du Laney, Brooklyn, N.Y.	ca.1866	after 1875
Love	Love Mfg. Co., Pittsburgh, Pa.	1885	after 1886
Lyon	Lyon Sewing Machine Co.	1879	ca.1880
Macauley	Thos. A. Macauley Mfg., New York, N.Y.	before 1879	—
Manhattan	Manhattan Sewing Machine Co.	ca.1868	ca.1880

70

Sewing Machine	Manufacturer or Company	First Made or Earliest Record	Discontinued or Last Record
McKay	McKay Sewing Machine Assoc.	1870	1876
McLean and Hooper	B. W. Lacy & Co., Philadelphia, Pa. (*see* Centennial)	ca.1869	ca.1873
Meyers	J. M. Meyers	1859	—
Miller's Patent	——	1853	—
Monitor (fig. 88)	Shaw & Clark Sewing Machine Co., Biddeford, Me.	1860	1864
Moore	Moore Sewing Machine Co.	ca.1860	—
Morey & Johnson (fig. 18)	Safford & Williams Makers, Boston, Mass.	1849	ca.1851
Morrison	Morrison, Wilkinson & Co., Hartford, Conn.	1881	—
Mower	——	ca.1863	—
National	Johnson, Clark & Co., Orange, Mass.	1874	—
National (also sold under distributor's name)	National Sewing Machine Co. (consolidation of the June and Eldredge Companies), Belvidere, Ill.	1890	1953
Ne Plus Ultra (fig. 110)	O. L. Reynolds Manufacturing Co., Dover, N.H.	1857	—
Nettleton & Raymond (fig. 111)	Nettleton & Raymond, Brattleboro, Vt.	ca.1857	—
New England (figs. 112, 113)	Charles Raymond (also by: Grout & White, Orange, Mass.; William Grout, Winchendon, Mass.; and J. G. Folsom, Winchendon, Mass.)	ca.1859 1862 1863 1865	1866 1863 — 1865
Newell	——	1881	—
New Fairbanks	J. H. Drew & Co. Thomas M. Cochrane Co., Belleville, Ill.	1878 1880	1880 —
New Home	New Home Sewing Machine Co., Orange, Mass. (in 1928 became affiliated with Free Sewing Machine Co.)	1876	*
New York	——, New York, N.Y.	ca.1855	ca.1855
New York Shuttle	N.Y. Sewing Machine Co., New York, N.Y. (later Demorest Mfg. Co.)	before 1880	1882

---

*Still in existence.

Sewing Machine	Manufacturer or Company	First Made or Earliest Record	Discontinued or Last Record
Noble	Noble Sewing Machine Co., Erie, Pa.	before 1881	after 1886
Novelty	C. A. French, Boston, Mass.	1869	—
Old Dominion	Old Dominion Sewing Machine Co., Richmond, Va.	ca.1858	1860
Pardox	——	ca.1865	—
Parham	Parham Sewing Machine Co., Philadelphia, Pa.	ca.1869	ca.1871
Parker	Charles Parker Co., Meriden, Conn. Later Parker Sewing Machine Co.	before 1860	after 1865
Pearl	—— Bennett	ca.1859	—
Philadelphia	Philadelphia Sewing Machine Co., Philadelphia, Pa.	ca.1872	ca.1881
Post Combination	Post Combination Sewing Machine Co., Washington, D.C.	before 1885	after 1886
Pratt's Patent (fig. 114) Later Ladies Companion	——	1857	ca.1858
Queen	Dauntless Mfg Co., Norwalk, Ohio	ca.1881	—
Quaker City (fig. 116)	Quaker City Sewing Machine Co., Philadelphia, Pa.	1859	ca.1861
Remington Empire Later Remington	Remington Empire Sewing Machine Co.	1870	1872
	E. Remington & Sons, Philadelphia, Pa.	1873	ca.1894
Robertson (dolphin & cherub) (figs. 40, 41)	T. W. Robertson, New York, N.Y.	1855	after 1860
Robinson	F. R. Robinson, Boston, Mass.	1853	ca.1855
Robinson's patent sewing machine with Roper's improvement (fig. 117) Later Robinson and Roper (fig. 118)	Howard & Davis, Boston, Mass. same	1855 1856	before 1860
Royal St. John (formerly St. John)	Royal Sewing Machine Co., Springfield, Ohio (later Free Co.)	ca.1883	1898
Ruddick	——	ca.1860	—
Secor	Secor Machine Co., Bridgeport, Conn.	1870	1876
Sewing Shears (Hendrick's patent) (fig. 43)	Nettleton & Raymond, Bristol, Conn.	ca.1859	—
Sewing Shears	American Hand Sewing Machine Co., Bridgeport, Conn.	ca.1884	ca.1900

Sewing Machine	Manufacturer or Company	First Made or Earliest Record	Discontinued or Last Record
Shaw & Clark Running Stitch Machine (fig. 53) Chainstitch Machine (fig. 119)	Shaw & Clark Co., Biddeford, Me.	ca. 1857	1866
Chainstitch Machine (fig. 120)	Shaw & Clark Co., Chicopee Falls, Mass.	1867	1868
	Chicopee Sewing Machine Co., Chicopee Falls, Mass.	1868	ca. 1869
Sigwalt	Sigwalt Sewing Machine Co., Chicago, Ill.	ca.1879	—
Singer (figs. 28, 29, 30, 32, 33, 121, 122)	I. M. Singer & Co. (later Singer Mfg. Co.), Moved from Boston to New York to Elizabethport, N.J. (factory).	1851	*
Springfield	Springfield Sewing Machine Co., Springfield, Mass.	1880	—
Standard (chainstitch) (fig. 123)	———	1870	—
Standard (shuttle)	Standard Shuttle Sewing Machine Co., New York, N.Y.	1874	ca.1881
Standard	Standard Sewing Machine Co., Cleveland, Ohio (acquired by Singer Co.)	1884	ca.1930
Stewart	Henry Stewart & Co., N.Y., N.Y.	1874	1880
Later New Stewart	Stewart Mfg Co.	1880	ca.1883
St. John (later Royal St. John)	St. John Sewing Machine Co., Springfield, O.	1870	ca.1883
Taggart & Farr (figs. 124, 125)	Taggart & Farr, Philadelphia, Pa.	1858	—
Thompson	C. F. Thompson Co.	1871	1871
	T. C. Thompson, Ithaca, N.Y.	ca.1854	—
Union	Johnson, Clark & Co., Orange, Mass.	1876	—
Victor	Finkle & Lyon Mfg. Co.	1867	ca.1872
	Victor Sewing Machine Co., Middletown, Conn.	ca.1872	ca.1890
Wardwell	Wardwell Mfg. Co., St. Louis, Mo.	ca.1876	1890
Watson (fig. 126)	Jones & Lee	1850	ca.1853
	Watson & Wooster, Bristol, Conn.	ca.1853	ca.1860
Waterbury	Waterbury Co., Waterbury, Conn.	1853	ca.1860
Weed	T. E. Weed & Co. (became Whitney & Lyons)	1854	—

*Still in existence.

Sewing Machine	Manufacturer or Company	First Made or Earliest Record	Discontinued or Last Record
Weed	Weed Sewing Machine Co. (reorganized from Whitney & Lyons), Hartford, Conn.	1865	—
Family Favorite		1867	—
Manu. Favorite		1868	—
General Favorite		1872	—
Hartford		1881	ca.1900
Wesson	Farmer & Gardner Manufacturing Co.	1879	1880
	D. B. Wesson Sewing Machine Co., Springfield, Mass.	1880	—
West & Willson (fig. 127)	West & Willson Co., Elyria, Ohio	1858	—
A. B. Wilson (fig. 23)	E. E. Lee & Co., New York, N.Y.	1851	1852
A. B. Wilson's patent seaming lathe Later Wheeler and Wilson (fig. 26, 27, 128, 129)	Wheeler, Wilson, Co., Watertown, N.Y.	late 1851	1856
	Wheeler & Wilson Mfg. Co., Bridgeport, Conn.	1856	1905
	Singer Co., Bridgeport, Conn.	1905	1907
White (fig. 130)	White Sewing Machine Co., Cleveland, Ohio	1876	*
Whitehill	Whitehill Mfg Co., Milwaukee, Wis.	ca.1875	after 1886
Whitney	Whitney Sewing Machine Co., Paterson, N.J.	ca.1872	ca.1880
Whitney & Lyons	Whitney & Lyons (a machine based on the 1854 patent of T. E. Weed)	ca.1859	ca.1865
Wickersham	Butterfield & Stevens Mfg. Co., Boston, Mass.	1853	—
Willcox & Gibbs (figs. 39, 131)	Willcox & Gibbs Sewing Machine Co., New York, N.Y.	1857	*
Williams & Orvis	Williams & Orvis Sewing Machine Co., Boston, Mass.	ca.1859	after 1860
Wilson (fig. 89) (see Buckeye)	Wilson (W.G.) Sewing Machine Co., Cleveland, Ohio	ca.1867	after 1885
Windsor (one thread)	Vermont Arms Co., Windsor, Vt.	1856	1858
Windsor (see Clark's Revolving Looper)	Lamson, Goodnow & Yale, Windsor, Vt.	1859	1861

*Still in existence.

Sewing Machine	Manufacturer or Company	First Made or Earliest Record	Discontinued or Last Record
Name Unknown	John W. Beane	1853	—
"	Henry Brind	1860	—
"	Garfield Sewing Machine Co.	1881	—
"	Geneva Sewing Machine Co.	1880	—
"	Gove & Howard	1855	—
"	Charles W. Howland, Wilmington, Del.	ca.1860	—
"	Miles Greenwood & Co., Cincinnati, Ohio	ca.1861	—
"	Hood, Batelle & Co.	1854	1854
"	Wells & Haynes	1854	1854
"	Wilson H. Smith, Birmingham, Conn.	ca.1860	—

Figure 68.—AMERICAN BUTTONHOLE, Overseaming & Sewing Machine of about 1870. Using serial numbers, these machines can be dated approximately as follows: 1—7792, 1869; 7793—22366, 1870; 22367—42488, 1871; 42489—61419, 1872; 61420—75602, 1873; 75603—89132, 1874; 89133—103539, 1875; and 103540—121477, 1876. Figures are not available for the years from 1877 to 1886. (Smithsonian photo 46953–E.)

Figure 69.—(NEW) AMERICAN SEWING MACHINE of about 1874. Illustration is from a contemporary advertising brochure. (Smithsonian photo 33507.)

75

Figure 70.—AMERICAN MAGNETIC SEWING MACHINE, 1854. Machines of this type were manufactured for only two years under the patent of Thomas C. Thompson, March 29, 1853, and later under the patents of Samuel J. Parker, April 11, 1854, and Simon Coon, May 9, 1854. On September 30, 1853, Elias Howe listed receipts of $1000 from the American Magnetic Sewing Machine Co. for patent infringement. The machines manufactured after that date carry the Howe name and 1846 patent date to show proper licensing. Judging by Howe's usual license fee of $25 per machine, about 40 machines were manufactured prior to September 1853. The company was reported to have made about 600 machines in 1854 before it went out of business. The only American Magnetic machine known to be in eixstence is in the collection of the Northern Indiana Historical Society at South Bend, Indiana. (*Photo courtesy of the Northern Indiana Historical Society.*)

Figure 71.—ATLANTIC SEWING MACHINE, 1869. This machine is typical of the many varieties manufactured for a very short time in the 1860s and 1870s. It is about the size of the average hand-turned variety, 8 by 10 inches, but lighter in weight. The frame design was the patent of L. Porter, May 11, 1869, and the mechanism was patented by Alonzo Porter, February 8, 1870. The latter patent model bears the painted legend "Atlantic" and is stamped "Aprl 1, 69," indicating that it was probably already in commercial production. This date possibly may refer also to L. Porter's design patent, since actual date of issue was usually later than date of application. (Smithsonian photo 48329–A.)

Figure 72.—A. Bartholf sewing machine, 1853. Abraham Bartholf of New York began manufacturing Blodgett & Lerow machines (see fig. 20) about 1850; the style and mechanics of these machines, however, were primarily those of the Blodgett & Lerow patent as manufactured by O. C. Phelps and Goddard, Rice & Co. For this reason they are considered Blodgett & Lerow—not Bartholf—machines.

The true Bartholf machine evolved when the manufacturer substituted Howe's reciprocating shuttle for the rotary shuttle of the Blodgett & Lerow machine, continuing to manufacture the machine in his own adapted style. Bartholf manufactured reciprocating-shuttle machines as early as 1853, and his was one of the first companies licensed by Howe.

All Bartholf machines licensed under Howe's patent carry the Howe name and patent date. They are sometimes mistakenly referred to as Howe machines, but they are no more Howe machines than those manufactured by Wheeler & Wilson, Singer, or many others.

On April 6, 1858, Bartholf was granted a patent for an improvement of the shuttle carrier. He continued to manu-facture sewing machines under the name "Bartholf Sewing Machine Co." until about 1865.

Using serial numbers, Bartholf machines can be dated approximately as follows:

Serial Number	Year	Serial Number	Year
1—20	1850	291—321	1855
21—50	1851	322—356	1856
51—100	1852	357—387	1857
101—235	1853	388—590	1858
236—290	1854	591—1337	1859

No record of the number of machines produced by Bartholf after 1859 is available.

The Bartholf machine illustrated bears the serial number 128 and the inscription "A. Bartholf Manfr, NY—Patented Sept. 1846 E. Howe, Jr." This machine is in the collection of the Baltimore County Historical Society. Note the close similarity between it and the 1850 Blodgett & Lerow machine manufactured by Bartholf. (*Photo courtesy of the Baltimore County Historical Society.*)

Figure 73.—BARTLETT SEWING MACHINE, 1867. The Bartlett machine was first manufactured in 1866 under the January 31, 1865, and October 10, 1865, patents of Joseph W. Bartlett. The machines were made by Goodspeed & Wyman for the Bartlett Co. and were so marked. The inventor received another patent on April 7, 1868, and later machines carry this third date also. Although the first few hundred machines did not bear the dates of patents held by the "Combination," before the end of the first year of production Bartlett was paying royalties. He continued to manufacture sewing machines until the early seventies when he converted to the manufacturing of street lamps.

Using serial numbers, Bartlett's machines can be dated approximately as follows: 1—1000, 1866; 1001—3126, 1867; 3127—?, 1868. There is no record of serial numbers for the succeeding years. (Smithsonian photo 45524–G.)

Figure 74.—BARTRAM & FANTON SEWING MACHINE, 1867. These machines were first manufactured in 1867 under the patents of W. B. Bartram, notably his patent of January 1, 1867. Three machines were exhibited at The Eleventh Exhibition of the Massachusetts Charitable Mechanics Association in 1869 where they were awarded a bronze medal. They were compared favorably to the Willcox & Gibbs machine (see fig. 39), which they resembled. Bartram received additional patents in the early seventies and also manufactured lock-stitch machines.

Using serial numbers, machines may be approximately dated as follows: 1—2958, 1867; 2959—3958, 1868; 3959—4958, 1869; 4959—5958, 1870; 5959—6962, 1871; 6963—7961, 1872; 7962—8961, 1873; and 8962—9211, 1874. (Smithsonian photo P63198.)

Figure 75.—BECKWITH SEWING MACHINE, 1871. Among the inventors whose patent claims were "to produce a cheap and effective sewing machine" was William G. Beckwith. His machine was first manufactured by Barlow & Son, and it realized considerable success in the few years of its production. The earliest model was operated like a pair of scissors or with a cord and ring as illustrated. Beckwith later added a hand crank. The machine was purchased in Crewe, Cheshire, England; it is stamped "Pat. April 18, 71 by Wm. G. Beckwith, Foreign Pats. Secured, Barlow & Son Manuf. N.Y., [serial number] 706." By 1874 the machines were marked "Beckwith S.M. Co." and two 1872 patent dates were added.

Using serial numbers, machines may be dated approximately as follows: 1—3500, 1871; 3501—7500, 1872; 7501—12500, 8173; 12501—18000, 1874; 18001—23000, 1875; 23001—?, 1876. (Smithsonian photo 46953–C.)

Figure 76.—BOUDOIR SEWING MACHINE, 1858. This machine, a single-thread, chainstitch model was based on the patents of Daniel Harris, dated June 9, 1857, June 16, 1857, and October 5, 1858. Manufactured primarily by Bennett in Chicago in

1859, it also may have been produced in the East, although no manufacturer's name can be found.

In 1860, the Boudoir, also called Harris's Patent sewing machine, was exhibited at the Massachusetts Charitable Mechanics Association Exhibition where it won a silver medal for "its combination of parts, its beauty and simplicity, together with its ease of operation." At this time the machine was described as making a "double lock stitch" (another name for the double chainstitch). It was also described as having been before the public for some time and combining "the improvements of others for which the parties pay license." The machine head was positioned on the stand similarly to that of the West & Willson (fig. 127) and stitched from left to right.

It is not known exactly how many of these machines were made or how long they were in vogue. Manufacture, although probably ceasing in the 1860s, is known to have been discontinued before 1881, when a list of obsolete sewing machines was published in *The Sewing Machine News*. (Smithsonian photo P63199.)

Figure 77.—(New) Buckeye sewing machine of about 1875. The Buckeye machine was one of several manufactured by W. G. Wilson of Cleveland, Ohio. It was licensed under Johnson's extended patent of April 18, 1867. Although it was small and hand turned, it used two threads and a shuttle to form a lockstitch. The machine was sufficiently popular for Wilson to introduce an improved model in the early 1870s, which he called the New Buckeye. W. G. Wilson continued to manufacture sewing machines until about the mid-eighties, although the Buckeye machines were discontinued in the seventies. (Smithsonian photo 45524–A.)

Figure 78.—Centennial sewing machine, 1876. The Centennial machine was basically a McLean and Hooper sewing machine which was renamed to take advantage of the coming Centennial celebration. It was based on the patents of J. N. McLean, March 30, 1869, and August 2, 1870, and made a two-thread chainstitch. Only about five hundred Centennial machines were manufactured in 1873, but by 1876 over three thousand had been constructed. The machines were advertised on white circulars which were printed in red and blue, and engraved with two women sewing, one by hand, labeled "Sewing in 1776," and one at a Centennial sewing machine, labeled "Sewing in 1876." There is no record that the machines were made after 1876. (Smithsonian photo 48216–T.)

Figure 79.—CLARK's Revolving-Looper double-thread sewing machine, 1860. This machine was manufactured by Lamson, Goodnow, & Yale of Windsor, Vermont. It was an attempt to improve on the combined ideas of the Grover and Baker machine, the Nettleton & Raymond machine, and the earlier single-thread Windsor machine. The improvements were made and patented by Edwin Clark on December 6, 1859. Widely advertised, the machines sold for $35 with a foot-power table. They could also be operated by hand. Over three thousand were manufactured and sold, and preparations were being made to continue manufacture of the earlier single-thread Windsor, originally made by the company's predecessor, Vermont Arms Co., when the Civil War broke out. A flood of arms orders arrived, and the sewing-machine manufacture was discontinued early in the summer of 1861. The sewing-machine equipment and business was sold to Grout & White of Massachusetts. (Smithsonian photo 48216.)

Figure 80.—Du LANEY SEWING MACHINE of about 1872. Most of the small, simple, chainstitch sewing machines of this period were constructed so that they could either be turned by hand or set into a treadle-powered table. Du Laney's Little Monitor, manufactured for only a few years, was based on the patents of G.L. Du Laney, July 3, 1866, and May 2, 1871. It was a two-thread, chainstitch machine powered only by a foot treadle. By simple adjustment, the machine could also make the cablestitch and the lockstitch. (Smithsonian photo 48221–C.)

Figure 81.—Eureka sewing machine, 1859. An example of the many short-lived types of which no written record can be found, this particular machine was used as a patent model for certain minor improvements in 1859. It has the name "Eureka" painted on the top and the following inscription incised on the baster plate: "Eureka Shuttle S. M. Co. 469 Broadway, N.Y." Although it is a shuttle machine, it carries no patent dates and was not included in the Howe royalty records. Neither is it listed in the obsolescence list published in 1881. The company probably could not pay its royalty fees and was forced out of business almost immediately. If this machine had not been used as a patent model, no record of the company's existence might remain. It should be noted that as in most shuttle machines the head was meant to be set into a treadle-powered table. Since most tables are very similar, they are not required for identification. (Smithsonian photo 48328–C.)

Figure 82.—M. Finkle sewing machine, 1857. The M. Finkle machines were manufactured in 1856 and 1857. Sometime before or about 1859, the inventor, Milton Finkle, formed a partnership and the machines were subsequently called M. Finkle & Lyon and later simply Finkle & Lyon. In 1859 the machine was awarded a silver medal by the American Institute for producing superior manufacturing and family lockstitch sewing machines. It also won a silver medal in Boston in 1860 at the Massachusetts Charitable Mechanics Association Exhibition. Although the name of the machine was changed to Victor in 1867, the company name remained Finkle & Lyon until about 1872 when it was changed to Victor also. Victor machines were manufactured until about 1890.

Machines can be dated by their serial number approximately as follows:

Serial Number	Year	Serial Number	Year
1—200	1856	13001—15490	1867
201—450	1857	15491—17490	1868
451—700	1858	17491—18830	1869
701—950	1859	18831—21250	1870
951—1500	1860	21251—28890	1871
1501—3000	1861	28891—40790	1872
3001—5000	1862	40791—48240	1873
5001—7000	1863	48241—53530	1874
7001—9000	1864	53531—59635	1875
9001—11000	1865	59636—65385	1876
11001—13000	1866		

No estimates are available for the years 1877 to 1890. (Smithsonian photo 48216–A.)

84

Figure 83.—FLORENCE SEWING MACHINE. The Florence machine was based on the patents of Leander W. Langdon, whose first patent was obtained in 1855. Langdon sewing machines were manufactured by the inventor for a few years. It was his patent of March 20, 1860, that was the immediate forerunner of the Florence machine, whose name was derived from the city of manufacture, Florence, Massachusetts. The Howe royalty records of 1860 listed the Florence Sewing Machine Co. as one that took out a license that year. Langdon's patent of July 14, 1863, was incorporated into the machines manufactured after that date; however, the date is always incorrectly stamped "July 18, 1863." In 1865, the machine won a silver medal at the Tenth Exhibition of the Massachusetts Charitable Mechanics Association.

Over 100,000 Florence machines were manufactured by 1870. About 1880 the company changed the name of the machine to Crown. Improvements led to the name New Crown by 1885. About this time the right to use the name Florence for a sewing machine was purchased by a midwestern firm for an entirely different machine. In 1885 the Florence company began to manufacture lamp stoves and heating stoves and shortly thereafter they discontinued the manufacture of sewing machines.

Using the serial numbers, Florence machines can be dated approximately as follows:

Serial Number	Year	Serial Number	Year
1—500	1860	82535—96195	1869
501—2000	1861	96196—113855	1870
2001—8000	1862	113856—129802	1871
8001—20000	1863	129803—145592	1872
20001—35000	1864	145593—154555	1873
35001—50000	1865	154556—160072	1874
50001—60000	1866	160073—164964	1875
60001—70534	1867	164965—167942	1876
70535—82534	1868		

No record of the number of machines produced each year between 1877 and 1885 is available.

The machine shown here, serial number 49131, was manufactured in 1865. It is stamped with the following patent dates: "Oct. 30, 1855, Mar. 20, 1860, Jan. 22, 1861, and July 18, 1863" and the Wilson patent date "Nov. 12, 1850." The machines from 1860–1863 are marked with the early Langdon patents, excluding the 1863 one, and they have the additional patent dates of Howe and others: "Sept. 10, 1846, Nov. 12, 1850, Aug. 12, 1851, May 30, 1854, Dec. 19, 1854, Nov. 4, 1856." (Smithsonian photo 45572–A.)

Figure 84.—GLOBE SEWING MACHINE. J. G. Folsom received two design patents in 1864, one on March 1 for a spool holder and one on May 17 for the basic style of the machine. Also in the same year, he was awarded a mechanical patent for an adjustment in the lower looper that would accommodate a change in needle size. Using these patents, he manufactured a single-thread, chainstitch machine, the Globe. Folsom also exhibited his machines at the Tenth Exhibition of the Massachusetts Charitable Mechanics Association in 1865. The Globe attracted particular attention and was awarded a silver medal.

In 1866 Folsom devised a new treadle attachment for hand-operated machines; the invention was featured in *Scientific American*, volume 14, number 17, with a Globe machine. Folsom again exhibited at the Massachusetts Mechanics exhibition in 1869. In addition to an improved single-thread Globe, he also showed a double-thread, elastic-stitch (double chainstitch) machine for which he received a silver medal.

Folsom machines were manufactured until 1871; 280 machines were manufactured in that year.

The Globe sewing machine illustrated is stamped "J. G. Folsom, Maker, Winchendon, Mass. Patented April 28, 1863 [Ketchum's patent], Mar. 1, 1864. May 17, 1864." The machine was manufactured before November 1864 or it would include the patent for the lower loop adjustment. (Smithsonian photo 48216–H.)

NOTE: At least five sewing machines, those in figures 84 through 89, are similar enough in appearance to cause some confusion, because their basic design stems from a short pillar.

Figure 86.—EMPIRE SEWING MACHINE, late 1860s. Although an Empire Sewing Machine Co. existed in New York in the 1860s (the predecessor of the Remington-Empire Co.), it is not known whether this machine was manufactured by that same company, which was primarily concerned with producing shuttle machines. This chainstitch machine is marked "Empire Co., Patented April 23, 1863," the date referring again to Ketchum's patent. It is very similar to Folsom's Globe, except that it has claw feet rather than a closed base; the painted designs on the base of both are almost identical to those on the Monitor. Its spool holder, mounted in reverse, is a crude imitation of the Folsom patent. The Empire machines were probably manufactured about the same time as the Wilson machine. (*Photo courtesy of The Henry Ford Museum and Greenfield Village, Dearborn, Michigan.*)

Figure 85.—GLOBE SEWING MACHINE with treadle attachment as illustrated in *Scientific American*, April 21, 1866. (Smithsonian photo 48221–A.)

Figure 87.—ATWATER SEWING MACHINE, 1858. Atwater machines, based on the patent of B. Atwater, issued May 5, 1857, were manufactured from 1857 to about 1860. The machine illustrated, which is designed to be operated by a hand-turned wheel, has an upper forked dog feed, and its horizontally supported spool is directly over the stitching area. Like the others, it has a striated pillar and claw feet. The manufacturer is unknown. (Smithsonian photo P63200.)

Figure 88.—Monitor sewing machine, 1860–1866. The Monitor machines of this style were not marked by their manufacturers, Shaw & Clark of Biddeford, Maine. Later the company was forced by the "Combination" to pay a royalty, so it changed the style and began marking its machines with the company name and patent dates (see fig. 119 for copy of seal). The Monitor, which employed the conventional vertical spindle to hold the spool of thread, had a top feed in the form of a walking presser. Its striated pillar was similar to that of the Atwater machine, and both featured the same claw feet and urn-like top. Unlike the Atwater, however, the Monitor had a double drive from the hand-turned wheel, which was grooved for operation with belt and treadle. (Smithsonian photo 33458.)

Figure 89.—Wilson sewing machine, late 1860s to early 1870s. In addition to the Buckeye (see fig. 77), W. G. Wilson manufactured several other styles of sewing machines. This one, a combination of the varying styles of the earlier pillar machine has even duplicated the general style of the spool holder patented by Folsom. The pillar is not striated, but the machine does repeat the claw feet of the Atwater and Monitor machines. Wilson machines are usually marked "Wilson Sewing Mach. Manuf'g Co. Cleveland, Ohio, Ketchum's Patent April 28, 1863." The latter name and/or patent date are found on many of the machines of this general construction. The patent is that issued to Stephen C. Ketchum for his method of converting rotary motion into reciprocal motion. (*Photo courtesy of The Henry Ford Museum and Greenfield Village, Dearborn, Michigan.*)

Figure 90.—GRANT BROTHERS SEWING MACHINE, 1867. This machine was one of several styles that utilized Raymond's 1861 patented chainstitch method. This machine, however, used an under feed rather than a top feed.

Neither a name nor a date appears on the machine. In the June 25, 1907, issue of the *Sewing Machine Times* it was called the Common Sense machine, but detailed research has turned up no evidence to substantiate this name. However, a dated brochure advertising the Grant Brothers machine and showing a model identical to that illustrated in the *Sewing Machine Times* has been found. The brochure states that the machine made an elastic lockstitch; this was not a true lockstitch, however, but was in fact a simple chainstitch.

Grant Brothers sold their machine, which had silver-plated mountings, for $18; the price included hemmer, Barnum's self-sewer, oilcan, screwdriver, clamp, gauge, and four silver needles. An additional charge of $12 was made for a table and treadle. Compared to other chainstitch machines the price was high, and the company was short-lived. (Smithsonian photo 60794–E.)

Figure 91.—GREENMAN AND TRUE SEWING MACHINE. This lockstitch machine based on S. H. Roper's patent of 1857 was manufactured at Norwich, Connecticut, from 1859 to 1861 by Cyrus B. True, the inventor, and Jared F. Greenman, True's financial partner. Licensed by the "Combination" and carrying the Howe patent date, the machine had obvious merit: it was strong, well made—a good family machine. Exhibited at the Ninth Exhibition of the Massachusetts Charitable Mechanics Association in September 1860, it received a bronze medal. (At this time the company was listed as Morse and True—the inventor had obviously taken on a second financial backer.) Unfortunately, the best market for the machine lay in the South, and the outbreak of the Civil War made collections impossible. This greatly retarded business and finally drove the firm into bankruptcy. In all, it is doubtful that more than one thousand machines were produced in the three years of manufacture.

The machine illustrated is marked "Greenman and True" and bears the serial number 402; it was probably manufactured early in 1860. (Smithsonian photo 48216–N.)

Figure 92.—GROVER AND BAKER SEWING MACHINE. The Grover and Baker machine was one of the more popular machines from the 1850s until the early 1870s. The company produced iron-frame machines, fine cabinet models, and portables (figs. 35 and 36). Their machines may be dated by serial number approximately as follows:

Serial Number	Year	Serial Number	Year
1—500	1851	5039—7000	1856
501—1000	1852	7001—10681	1857
1001—1658	1853	10682—15752	1858
1659—3893	1854	15753—26033	1859
3894—5038	1855	26034—44869	1860

Serial Number	Year	Serial Number	Year
44870—63705	1861	225887—261004	1869
63706—82641	1862	261005—338407	1870
82642—101477	1863	338408—389246	1871
101478—120313	1864	389247—441257	1872
120314—139148	1865	441258—477437	1873
139149—157886	1866	477438—497438	1874
157887—190886	1867	497439—512439	1875
190887—225886	1868		

(Smithsonian photo 45513–B, an engraving of a Grover and Baker sewing machine from an advertising brochure of about 1870.)

Figure 93.—HANCOCK SEWING MACHINE, 1867. One of the many inventors who turned his talents to inventing and producing a mechanically simple and cheaper machine was Henry J. Hancock. His 1867 machine is only about six inches wide; it uses a tambour-type needle, pulling a loop of thread from below the stitching surface. (Smithsonian photo P63197.)

Figure 94.—HANCOCK SEWING MACHINE, 1868. Hancock in 1868 received both a design patent and a mechanical patent now using the eye-pointed needle and a hook to form the chainstitch. The design was an open framework circle with a mirror mounted in front of the table clamp. The purpose of the designated "looking glass" was decorative only. The Hancock machines were only manufactured for a few years. They measure 10½ inches in width, slightly larger than the earlier machine. (Smithsonian photo 48328–M.)

Figure 95.—[A.C.] HERRON'S PATENT SEWING MACHINE, 1858. The manufacturer of this machine is not known, but the machine was based on the patent of Abial C. Herron issued August 4, 1857. All the machines carry a small heart-shaped plate just above the needle descent bearing the patentee's name and the patent date. The patent covered an improvement in the method of making the chainstitch. The machines were provided with a hand crank, but were also meant to be operated by a belt and treadle. No records of the extent of manufacture of this machine have been found. This machine head measures 14 inches in width, about standard size. (Smithsonian photo 48329–J.)

Figure 96.—A. B. HOWE SEWING MACHINE of about 1860. (Smithsonian photo 45525–C.)

Figures 96, 97, and 98.—THE HOWE MACHINES. It is difficult for many to believe that the stamped legend "Elias Howe patent, Sept. 10, 1846" does not certify that a machine is an original Howe. Although Elias Howe was granted a patent for the lockstitch machine in 1846, he did not establish a sewing-machine factory for about twenty years. Early in the 1850s and later through the "Combination," however, he licensed others to make machines using his patent. These machines bore that patent date for which a royalty was being paid.

Among his early licensees was his elder brother Amasa who organized the Howe Sewing Machine Co. in 1854. The Amasa Howe machines were very good ones, and in 1862 Amasa won the prize medal at the London International Exhibition. This immensely increased the popularity of the machine and Elias offered to join Amasa by building a large factory at Bridgeport, Connecticut, to fill the increasing demand for more machines. The machines produced at Bridgeport, however, although imitating the Amasa Howe machines, proved inferior in quality. Amasa found that, rather than helping his business reputation, his brother's efforts were hurting him, and he severed business relations with Elias.

Because of their brief association, the 1862 prize medal awarded to A. B. Howe was sometimes credited to Elias. The latter did receive awards for his patent, but never for his manufactured machines. When the two brothers dissolved their joint venture, Elias attempted to call his new company the Howe Sewing Machine Co., but Amasa's claim that this name had been his exclusive property for many years was upheld by the courts. Elias then omitted the word

"Sewing" and called his company simply the Howe Machine Co.

After Elias died in 1867, the company was run by his sons-in-law, the Stockwell brothers. To distinguish their machines from those of A. B. Howe, they marked each machine with a brass medallion picturing the head and flowing locks of Elias Howe. They also continued to advertise their machine as the "original" Howe. In about 1873, B. P. Howe, Amasa's son, sold the Howe Sewing Machine Co. to the Stockwell brothers, who continued to manufacture Howe machines until 1886.

The machines of the A. B. Howe Sewing Machine Co. may be dated by serial number approximately as follows:

Serial Number	Year	Serial Number	Year
1—60	1854	167—299	1857
61—113	1855	300—478	1858
114—166	1856	479—1399	1859

No figures are available for 1860–1870, but 20,051 machines were manufactured in 1871.

The machines of the [Elias] Howe Machine Co. are not believed to have begun with serial number 1, and no figures are available for 1865–1867. After that, the machines may be dated by serial number approximately as follows:

Serial Number	Year	Serial Number	Year
11,000—46,000	1868	446,011—536,010	1873
46,001—91,843	1869	536,011—571,010	1874
91,844—167,000	1870	571,011—596,010	1875
167,001—301,010	1871	596,011—705,304	1876
301,011—446,010	1872		

No figures are available for 1877–1886.

ILLUSTRATIONS

OF THE

GENUINE HOWE MACHINES.

None others are genuine.

CAUTION.—Before purchasing, see the name "HOWE" is on the arm of the Machine.

SOLE DEPÔT FOR GREAT BRITAIN: 8, LUDGATE STREET, ST. PAUL'S, LONDON.

Letter A Machine.

Letter B Machine.

Letter C Machine.

Letter D Machine.

Figure 97.—ADVERTISING BROCHURE distributed by E. Howe during the brothers' brief partnership; the machines are basically A. B. Howe machines, 1863. (Smithsonian photo 49373–A.)

92

Figure 98.—Howe (Stockwell brothers) machine, 1870. (Smithsonian photo 45572–E.)

Figure 99.—PATENT MODEL OF CHRISTOPHER HODGKINS, November 2, 1852, assigned to Nehemiah Hunt. (Smithsonian photo 34551.)

Figures 99, 100, and 101.—THE N. HUNT (later, in 1856, Hunt & Webster and finally in 1858 Ladd and Webster) sewing machine was based on the patents of Christopher Hodgkins, November 2, 1852, and May 9, 1854, both of which were assigned to Nehemiah Hunt. First manufactured in 1853, the machine, which closely resembled the Hodgkins' patent, won a silver medal at the exhibition of the Massachusetts Charitable Mechanics Association that same year.

In 1856 Hunt took a partner, and the company became Hunt & Webster. An interesting account of this company appeared as a feature article in *Ballou's Pictorial*, July 5, 1856, where it was reported that "the North American Shoe Company have over fifty of the latest improved machines, represented in these drawings [fig. 31], now running . . . ." The article also estimated that a 55-million dollar increase in shoe manufacturing in Massachusetts in

1855 was due to the sewing machine. In 1856 the Hunt & Webster machine again won a silver medal at the exhibition. Very late in 1858 the company became Ladd, Webster, & Co. and continued to manufacture both family and manufacturing sewing machines until the mid-1860s.

The approximate date of manufacture can be determined by serial number:

Serial Number	Year
1—100	1853
101—368	1854
369—442	1855
443—622	1856
623—1075	1857
1076—1565	1858
1566—3353	1859

No figures are available for the 1860s.

Figure 100.—Right: Hunt & Webster sewing machine of about 1855, serial number 414. (Smithsonian photo 48216–V.)

Figure 101.—Ladd, Webster & Co. sewing machine of about 1858, Boston, serial number 1497. (Smithsonian photo 46953.)

Figure 102.—IMPROVED COMMON SENSE sewing machine of about 1870. This machine is so very

similar to the New England machines in its feed, threading, looping mechanism, and in its general design, that it is sometimes mistaken for the earlier New England machines (see figs. 112 and 113).

Dating from the early 1870s, the Improved Common Sense machine is about 10 inches in width, two inches larger than the New England machine. The spool holder is similar to Folsom's patented design, but is less refined. A page from an advertising brochure of the period verifies the name of the machine, but does not identify the manufacturer.

There are no patent dates or identifying names or numbers on the machine illustrated. Although the Empire Co. also produced a machine of this style, their models are marked with their name and with Ketchum's patent date, April 23, 1863. Of the several styles of machine using the Raymond looper, this type seems to account for the largest volume manufactured, as evidenced by the proportionately higher number of examples still extant. (Smithsonian photo 48328–E.)

Figure 103.—JOHNSON SEWING MACHINE, 1857. Another of the all-but-forgotten manufacturers of the 1850s was Emery, Houghton & Co., who constructed the A.F. Johnson machines. Examination of existing machines indicates that they were manufactured in 1856 and 1857, and possibly a little longer. This one from 1857 bears the serial number 624, so we know that several hundred were manufactured. The head is ornately attractive, slightly reminiscent of Wheeler & Wilson models, and of standard size. (Smithsonian photo 48329–B.)

Figure 104.—"LADY" SEWING MACHINE of about 1859. The contemporary name of this machine is unknown. The unusual design of the head, or main support, is based in part on the design patent, number 216, of Isaac F. Baker, issued April 10, 1849, for a "new and useful design [,] for ornamenting furniture [,] called Cora Munro" who was a character in James Fenimore Cooper's *Last of the Mohicans*. The design shows a female figure wearing a riding dress and hat that is ornamented with a plume and a bow. Her right hand holds a riding stick and the left, her skirt. Trunks of trees and foliage complete the Baker design, which is known to have been used for girandoles of the period. A companion design was also patented by Baker, number 215, which is in the form of a man in military costume and is named "Major Heyward," for another character in *Last of the Mohicans*.

The sewing machines based on the "Cora Munro" design also use branch designs as the overhanging arms. A mother bird sits in the upper branch and descends to feed a young bird as the machine is in operation. The one illustrated was used as the machine submitted with a request for patent by George Hensel of New York City for which patent 24,737 was issued on July 12, 1859. Since Hensel's patent application was for an improvement in the feed, there was no need for the highly decorative head unless such a machine was commercially available. The patent specifications merely state that the head is "ornamented." Another sewing machine of this type was used as the patent model by Sidney Parker of Sing Sing, New York, number 24,780, issued on the same date as the Hensel patent. Parker's patent also covered an improved feeding mechanism. In the patent description, however, the inventor states that "the general form of the machine is not unlike others now in use." By this he might have meant in the design, or possibly in the basic structural form. Other than the two machines described, no other examples are known to have survived, but "Lady" or "Cora Munro" sewing machines were manufactured. (Smithsonian photo 45506–D.)

Figure 105.—LANDFEAR'S PATENT SEWING MACHINE of about 1857. Another of the many machines that, except for isolated examples, have almost completely disappeared from the records is Landfear's machine. Fortunately, this manufacturer marked his machine—where many did not—stamping it: "Landfear's patent – Decr 1856, No. 262, W. H. Johnson's Patent Feb. 26th 1856, Manfrd by Parkers, Snow, Brooks & Co., West Meriden, Conn." (There was a Parker sewing machine manufactured by the Charles Parker Co. of Meriden, but his machine was a double-thread chainstitch machine and was licensed by the "Combination." The Landfear machine may have been an earlier attempt by a predecessor or closely related company.)

The Landfear patent was for a shuttle machine, but it also included a mode for regulating stitch length. The name chosen for this machine may be incorrect, since the single-thread chainstitch mechanism is primarily that of W. H. Johnson, but since the Johnson patent also was used on other machines the name "Landfear" was assigned. The machine was probably another attempt to evade royalty payment to the "Combination."

The serial number 262 indicates that at least that many machines were manufactured, although this model is the only one known to be in existence. The support arm of the machine head is iron, cast as a vase of flowers and painted in natural colors. The paint on the head is original, but the table has been refinished, and the iron legs, which had rusted, have been repainted. (Smithsonian photo 48440–G.)

Figure 106.—Lathrop sewing machine of about 1873. These machines were manufactured by the Lathrop Combination Sewing Machine Co. under the patents of Lebbeus W. Lathrop of 1869, 1870, and 1873. The machine used two threads, both taken from spools; moreover, it produced not only the double chainstitch, but it was constructed to produce also a lockstitch and a combined "lock and chain stitch." The machine illustrated bears the serial number 31 and the patent dates of Grover & Baker, and Bachelder among others, in addition to the first two Lathrop patent dates. The company lasted only a few years as it is included in the 1881 list of manufacturers that had ceased to exist. (Smithsonian photo 46953–F.)

J. B. NICHOLS & CO.,

MANUFACTURERS OF

HOWES' IMPROVED PATENT

Figure 107.—ILLUSTRATION FROM A BROCHURE, marked in ink: "The National Portrait Gallery, 1855," Singer Archives. (Smithsonian photo 48091–E.)

Figures 107 and 108.—THE NICHOLS AND LEAVITT sewing machines. One of Elias Howe's earliest licensees was J. B. Nichols. His machine, manufactured at first with George Bliss and later alone as J. B. Nichols & Co., was called Howe's Improved Patent Sewing Machine. It was, however, no more a Howe machine than any of the others produced under the Howe patent.

In July 1855 Nichols went into partnership with Rufus Leavitt, and the company name changed to Nichols, Leavitt & Co. In 1857 it was changed again to Leavitt & Co., and finally in the mid-1860s to Leavitt Sewing Machine Co. By the 1870s, it was defunct.

The Nichols-Leavitt machines can be dated by their serial numbers approximately as follows:

Serial Number	Year	Company
1—28	1853	Nichols & Bliss
29—245	1854	J. B. Nichols & Co.

Serial Number	Year	Company
246—397	1855	J. B. Nichols & Co.—Nichols, Leavitt & Co.
398—632	1856	Nichols, Leavitt & Co.
633—827	1857	Leavitt & Co.
828—902	1858	"
903—1115	1859	"
1116—1436	1860	"
1437—1757	1861	"
1758—2077	1862	"
2078—2400	1863	"
2401—2900	1864	"
2901—3900	1865	Leavitt Sewing Machine Co.
3901—4900	1866	"
4901—5951	1867	"
5952—6951	1868	"
6952—7722	1969	"

There is no record that the company was in existence after 1869.

Figure 108.—Leavitt sewing machine of about 1868, serial number 6907. (Smithsonian photo 48328.)

Figure 109.—LESTER SEWING MACHINE of about 1858. The Lester machine was first manufactured by J. H. Lester in Brooklyn, New York. His machine was based on the patents of William Johnson, John Bradshaw and others but not on the patents held by the "Combination," although he had secured a license. When the Old Dominion Company applied for a license from the "Combination," Lester learned of this, went to Richmond, and arranged to combine his business with theirs. Since the Lester machine was the better one, it was agreed to cease the manufacture of the Old Dominion machines early in 1860 and in March the company name was changed to the Lester Mfg. Co. Late in 1860, George Sloat entered the company with his Elliptic machine; the name was changed again, this time to Union Sewing Machine Co. The manufacture of both sewing machines continued until the outbreak of the Civil War the following year, which brought a conversion to arms production. The manufacture of Lester machines was never resumed.

The machine illustrated was manufactured by J. H. Lester in Brooklyn; it bears the serial number 96. The number of Lester machines manufactured from 1858 through 1861 is not known, but it was probably less than 1,000. (Smithsonian photo P63359.)

Figure 110.—Ne Plus Ultra of about 1867. Another of the interesting hand-turned chainstitch machines of the late 1850s and 1860s was patented by O.L. Reynolds. The baster plates and the handle on the wheel are missing on this machine, but an interesting shield and draped-flag pattern is painted on the base.

Another machine of this type has the following inscription stamped on the baster plate: "Ne Plus Ultra, Patent Applied For, 174, O.L. Reynolds, Patentee & Manufacturer, Dover N.H." Reynold's patent model, March 30, 1858, bears the serial number 110, indicating that the machine illustrated here—which bears the serial number 26—was manufactured before the patent was obtained. (Smithsonian photo 48216–F.)

Figure 111.—NETTLETON & RAYMOND SEWING MACHINE. One of the most ornate of the early, small, hand-turned sewing machines was patented and manufactured by Willford H. Nettleton and Charles Raymond whose first patent was received on April 14, 1857. The patent model, believed to be a commercial machine, is beautifully silver-plated. Whether this was a special one-of-a-kind model, or whether the inventors tried to make a commercial success of a silver-plated machine is not known. The machine made a two-thread chainstitch, taking both threads from commercial spools. By October 1857, the inventors had received their second patent. This time the machine was brass and gilt—brighter, but less expensive. At the same time, Nettleton & Raymond began manufacturing sewing-shears machines under the patent of J. E. Hendricks.

By the latter half of 1858, Nettleton & Raymond had moved from Bristol, Connecticut, to Brattleboro, Vermont. The patented improvement of the two-thread chainstitch machine received that year was in the name of "Raymond, assignor to Nettleton," although the machines of this type bear neither name nor patent date. No record of the price for which they were sold has been found, but it would be fair to estimate that it was probably about $25. This style of machine was discontinued when the manufacture of the simpler, more profitable New England model began, a machine that Raymond had initiated just before the partners left Bristol. (Smithsonian photo 45505–E.)

Figure 112.—RAYMOND PATENT MODEL, March 9, 1858. (Smithsonian photo 32009–0.)

Figures 112 and 113.—NEW ENGLAND SEWING MACHINES. The small, hand-turned, sewing machines some of which were called Common Sense, were manufactured by at least three companies and possibly more. The earliest ones were those made by Nettleton & Raymond based on Charles Raymond's patent of March 9, 1858, which featured a hinged presser foot acting as the top feed. On July 30, 1861, Raymond received a patent for an improved looper; this date is found on all machines later manufactured by the inventor.

In 1858 Nettleton and Raymond had moved from Bristol, Connecticut, to Brattleboro, Vermont. Also in Brattleboro at this time were Thomas H. White and Samuel Barker, who were manufacturing a small machine called the Brattleboro. White left Vermont in 1862 and went to Massachusetts. There, in partnership with William Grout, he also began to manufacture New England machines; these were basically the same as the Raymond machines. After a short time, Grout left the partnership with White and moved to Winchendon, there continuing to make New England machines for approximately one more year. In 1865, J. G. Folsom of Winchendon exhibited a New England machine at the Tenth Exhibition of the Massachusetts Charitable Mechanics Association along with his Globe machine. Whether both machines were manufactured by him or whether he might have been exhibiting one of Grout's machines is not known.

There is no record that New England machines were manufactured after 1865. There is a great similarity between these machines and the Improved Common Sense sewing machines of the 1870s. It is believed that the name "Common Sense" was given by frugal New Englanders to several of the cheaper chainstitch machines of the 1860s.

Figure 113.—NEW ENGLAND sewing machine of about 1860, manufactured by Nettleton & Raymond; it bears the Raymond patent date of March 9, 1858. (Smithsonian photo 45505–G.)

Figure 114.—Pratt's second patent model, March 3, 1857, probably a commercial machine. (Smithsonian photo 48328–H.)

Figures 114 and 115.—Pratt's patent and the Ladies Companion sewing machine. The machines manufactured under the patents of Samuel F. Pratt were first sold in 1857 and 1858 as Pratt's patent. These machines carry the Pratt name and the patent dates "Feb. 3, 1857 Mar. 3;" the latter is an 1857 patent date also. In 1859 the Pratt machine was called the Ladies Companion and was so marked. It was also marked with the 1857 patent dates, the date February 16, 1858, and a serial number, and was stamped "Boston, Mass." Manufacture was discontinued after a few years.

Figure 115.—Ladies Companion, 1859. (*Photo courtesy of The Henry Ford Museum and Greenfield Village, Dearborn, Michigan.*)

Figure 116.—QUAKER CITY SEWING MACHINE. During the first decade of sewing-machine manufacture many types of handsome wooden cases were developed to house the mechanisms. Although such cases increased the total cost, they were greatly admired and were purchased whenever family funds permitted. The machine was based on the patents of William P. Uhlinger: a mechanical patent for a double chainstitch machine on August 17, 1858 (antedated May 8), and a patent for the casing on December 28, 1858. The machine head was lowered into the casing as the lid was brought forward and closed—an idea much ahead of its time.

This Quaker City machine, serial number 18, was purchased by Benjamin F. Meadows of Lafayette, Alabama, for $150 just prior to the Civil War. Relatively few machines of this type were manufactured, and the Quaker City Sewing Machine Co. existed for only a few years. Its apparent hope for a southern market was short-lived, and it was unable to compete either with the companies licensed under the "Combination" or with those producing less expensive machines. (Smithsonian photo 46953–A.)

Figure 117.—FROM AN ADVERTISING BROCHURE, marked in ink, "The National
Portrait Gallery, 1855," in the Singer Company's archives. The brochure
states "Howard & Davis, 34 Water Street, Boston, Massachusetts Sole
Manufacturers of Robinson's Patent Sewing Machine with Rope[r]'s
Improvements." (Smithsonian photo 48091–F.)

Figure 118.—SEWING MACHINE OF ABOUT 1856 with inscription "Howard &
Davis Makers, Boston, Mass. Robinson & Roper Pat. Dec. 10, 1850, Aug. 15,
1854"; the drive wheel and the circular stitching plate of this machine are
missing. (Smithsonian photo 48440–C.)

Figures 117 and 118.—ROBINSON AND ROPER sewing
machines, 1855–1856. This is one of the few machines producing a backstitch or half backstitch to realize any commercial success. Manufactured a very short time by Howard
& Davis, it was a short-thread machine, based on the
Frederick Robinson patent of December 10, 1850, and the
Samuel Roper patent of August 15, 1854. Roper produced
additional improvements for which he received a patent on
November 4, 1856. In the *Scientific American*, November 1,
1856, the new machine was discussed: "Robinson & Roper
exhibit their new improved sewing machines, which appear
to operate with great success. Two needles are employed,
the points of which are furnished with hooks that alternately
catch the thread and form the stitch. The finest kind of
cotton thread or silk can be used. The work appears well
done. Price $100."

Figure 119.—ILLUSTRATED PAGE in a Shaw & Clark advertising brochure, published in late 1864. (Smithsonian photo 61321.)

Figure 120.—SHAW & CLARK SEWING MACHINE (Page patent) of 1867, Chicopee Falls, Massachusetts. (Smithsonian photo 48216–L.)

Figures 119 and 120.—SHAW & CLARK SEWING MACHINES. In addition to the early style Monitor sewing machine sold by Shaw & Clark without a name or any identifying marks, the company continued to manufacture machines after a lawsuit with the "Combination" forced them to take out a license. They manufactured an adapted version of their Monitor and an entirely new design patented in 1861. Their machines were now marked with the company name and a list of patent dates including those of Howe, Wheeler and Wilson, Grover and Baker, and Singer and the Batchelder patent, together with their own design patents. In 1867 the company moved from Biddeford, Maine, to Chicopee Falls, Massachusetts. In the same year, they began manufacturing a machine of the design patented by T. C. Page. The company is believed to have become the Chicopee Sewing Machine Company which appeared the following year and remained in business only a very short time. One Chicopee sewing machine is in the Smithsonian collection.

Figure 121.—SINGER "Traverse Shuttle Machine—Letter A." (Smithsonian photo 58984.)

Figures 121 and 122.—SINGER SEWING MACHINES. From 1850 to 1858 the Singer company produced heavy manufacturing-type sewing machines similar to the patent model shown earlier (fig. 28). The first machine for family use, Singer's new "Family" sewing machine (fig. 33) was manufactured from 1858–1861. Their second-style family machine was called the "Traverse Shuttle Machine—Letter A;" it was manufactured from 1859 to 1865, when they introduced their third family machine and called it the "New Family" sewing machine. This style machine continued until about 1883 when the "Improved Family" machine appeared. In addition to the lockstitch machines, Singer also manufactured chainstitch machines, and many highly specialized manufacturing machines.

From 1857 through the 1880s, the Singer machines were marked with two serial numbers. It is possible that the numbers were related to the "Combination" royalties paid by the Singer company. Until about 1873 there was a difference of exactly 4,000 in the two numbers, thus one machine would be marked 12163 and directly below it would be marked 16163. From 1873 the last three digits of the two numbers continued to be the same but the lower number might be much lower in value than either number used in earlier years. The larger number is believed to have been a record of total production while the lower number may have referred to a machine of a particular style. The Singer company records can shed no light on the meaning of the top (or lower of the two) serial numbers. Generally, in the earlier machines, the difference in the two numbers will not affect the dating of a machine by more than one year.

Since dating by serial number can only be estimated, the two numbers do not add an appreciable variable prior to 1873. Only the larger number, however, should be considered in dating machines after 1873.

Serial Number	Year	Serial Number	Year
1—100	1850	99427—123058	1864
101—900	1851	123059—149399	1865
901—1711	1852	149400—180360	1866
1712—2521	1853	180361—223414	1867
2522—3400	1854	223415—283044	1868
3401—4283	1855	283045—369826	1869
4284—6847	1856	369827—497660	1870
6848—10477	1857	497661—678921	1871
10478—14071	1858	678922—898680	1872
14072—25024	1859	898681—1121125	1873
25025—43000	1860	1121126—1362805	1874
43001—61000	1861	1362806—1612658	1875
61001—79396	1862	1612659—1874975	1876
79397—99426	1863		

Since records of annual production from 1877 to the turn of the century are not complete, it is difficult to establish yearly approximations. Using the machines submitted as patent models, and thus known to have been manufactured before the date of deposit, however, has provided us with the following date guides. By 1877 there had been 2 million machines manufactured, 3 million by 1880, 4 million by 1882, 5 million by 1884, 6 million by 1886, 7 million by 1888, 8 million by 1889, 9 million by 1890, and 10 million by 1891.

Figure 122.—Singer "New Family" sewing machine. (Smithsonian photo 58987.)

Figure 123.—STANDARD SEWING MACHINE of about 1870. This chainstitch machine is believed to have been made by the company that later became the Standard Shuttle Sewing Machine Company, when they began manufacturing lockstitch machines about 1874. This machine is marked with the name, "Standard," and with the dates "Patented July 14, 1870, Patented Jan. 22, 1856, Dec. 9, 1856, Dec. 12, 1865." The dates refer to the reissue and extended reissue of the Bachelder and the A. B. Wilson patents. The number of chainstitch machines of this type that were manufactured is not known. (Smithsonian photo 45506–C.)

Figure 124.—TAGGART & FARR sewing machine, front view. (Smithsonian photo 48216–P.)

Figures 124 and 125.—TAGGART & FARR sewing machine, 1860. The Taggart & Farr is an almost forgotten machine. It was based on Chester Farr's patent of August 9, 1859. The machine, however, was in commercial production as early as 1858, the year the patent application was made. Using two threads—both taken directly from the spool—to form a chainstitch, the machine was operated basically by treadle but also by hand. The drive wheel is missing on this machine, but it would normally appear on the right.

The name and patent date were painted on the end of the machine. This was true of many other machines of this period, which is why so many go unidentified once the paint has become worn. Several thousand Taggart & Farr machines were manufactured, but the company is believed to have had a short life, for it was among those that had disappeared by 1881.

Figure 125.—TAGGART & FARR sewing machine, end view. (Smithsonian photo 48216–M.)

Figure 126.—WATSON SEWING MACHINE, 1856, illustrated in *Scientific American*, December 13, 1856. The earliest Watson machines were two-thread lockstitch machines, as described in the *Scientific American*, August 10, 1850. Although the magazine reported that the inventor had applied for a patent, the earliest lockstitch patent issued to William C. Watson was on March 11, 1856. A few of his machines were made in 1850, the article continued, "several of these machines are nearly finished . . . persons desirous of seeing them can be gratified by calling upon Messrs. Jones & Lee." A Watson machine was exhibited by Jones & Lee at the Sixth Exhibition of the Massachusetts Charitable Mechanics Association held in Boston in September 1850.

In 1853 a Watson machine was exhibited at the New York Industry of All Nations Exhibition, but this was a single-looping machine; Watson received a patent for this single-thread machine on November 25, 1856.

In the December 13, 1856, issue of *Scientific American* a machine called Watson's "Family" sewing machine was illustrated and described. It was a small machine (only 8 by 5 inches) manufactured by Watson & Wooster and selling for $10. References to the Watson single-thread machine occur as late as 1860, but no examples are known to have survived. (Smithsonian photo 48221–B.)

116

Figure 127.—West & Willson sewing machine of about 1859. The West & Willson machine, manufactured under the patent of H. B. West and H. F. Willson, enjoyed a very brief span of popularity. The patent covered the peculiar method of operating a spring-looper in combination with an eye-pointed needle to form a single chainstitch, but whether machines of this single-thread variety were manufactured is unknown. The machine illustrated here is a two-thread machine of basically the same description. It stitches from left to right and bears serial number 1544 and the inscription "West & Willson Co. patented June 29, 1858." (Smithsonian photo 49456–A.)

Figure 128.—Wheeler & Wilson sewing machine of about 1872. Serial
number 670974. (Smithsonian photo P63149–A.)

Figure 129.—WHEELER AND WILSON No. 8 sewing machine of about 1876. (Smithsonian photo 17663–C.)

Figures 128 and 129.—WHEELER AND WILSON sewing machines. The Wheeler and Wilson company was the largest manufacturer of sewing machines in the 1850s and the 1860s.

It began in 1851 as A. B. Wilson; from 1852 to 1856 it was the Wheeler, Wilson & Co., Watertown, Connecticut; and from 1856 to 1876, it was Wheeler & Wilson Mfg. Co., Bridgeport, Connecticut.

The style of the head changed very little during these years (see figs. 26 and 27). Both a table style with iron legs and a cabinet model were made: the head was usually mounted to stitch from left to right. In 1861, the company introduced the famous glass presser foot, patented on March 5 of that year by J. L. Hyde. The presser foot was made of metal but shaped like an open into which was slid a small glass plate, with a hole for the needle descent. The glass allowed the seamstress to observe the stitching and to produce very close-edge stitching. It remained a favorite of many women for years. In 1876, the new No. 8 machine was introduced and a new series of serial numbers was initiated. It is, therefore, imperative to know that the machine is one of the earlier style machines before using the following list of serial numbers to date the machines, approximately as follows:

Serial Number	Year	Serial Number	Year
1—200	1851	141100—181161	1864
201—650	1852	181161—220318	1865
651—1449	1853	220319—270450	1866
1450—2205	1854	270451—308505	1867
2206—3376	1855	308506—357856	1868
3377—5586	1856	357857—436722	1869
5587—10177	1857	436723—519930	1870
10178—18155	1858	519931—648456	1871
18156—39461	1859	648457—822545	1872
39462—64563	1860	822546—941735	1873
64564—83119	1861	941736—1034563	1874
83120—111321	1862	1034564—1318303	1875
111322—141099	1863	1138304—1247300	1876

Records of the second series of serial numbers dating from 1876 are not available.

Figure 130.—WHITE SEWING MACHINE. Although the White sewing machines date from 1876, Thomas H. White had been busy in the manufacture of sewing machines for many years prior to this. White is known to have been associated with Barker in the manufacture of the Brattleboro machine and later with Grout in producing one of the several New England machines. In 1866 he moved to Cleveland, Ohio, and began manufacturing machines for sale under special trade names through selling organizations. In 1876, the White Sewing Machine Company was formed and machines were sold under the White name.

The machine illustrated is a standard lockstitch machine, which would have been set into a sewing-machine table and operated by a treadle. The small handle was used to start

the wheel, and thus the stitching operation, in the forward direction. This machine bears the serial number 28241 and the following patents: "Mar. 14, 1876, May 2, 1876, Oct. 24, 1876, Jan. 16, 1877, Mar. 20, 1877, Mar. 27, 1877," which are primarily the patents of D'Arcy Porter and George W. Baker.

The machines of the 1870s may be dated approximately as follows:

Serial Number	Year
1—9000	1876
9000—27000	1877
27001—45000	1878
45001—63000	1879

(Smithsonian photo 48329–H.)

Figure 131.—WILLCOX AND GIBBS SEWING MACHINE, serial number 296572, of about 1878. From 1857 to the turn of the century, the style of the Willcox and Gibbs sewing machine changed very little (fig. 39). It was the most popular and the most reliable of the many chainstitch machines. In addition to the basic mechanical patents, Gibbs also patented the design of the sewing-machine head in 1860. In the specifications, he described it as an open ring set on a base or pedestal. The lower part of the open section supported the cloth plate. The design of the head, intentionally or not, formed a perfect letter G, the initial of the inventor. Later the machine head as a letter G was incorporated into the company's trademark. Additional patents were also granted to James Willcox for a leg and treadle design and to Charles Willcox for mechanical improvements.

It has not been possible to secure information on records of serial numbers from the late 1870s through the 1920s to aid in dating machines of that period. For the preceding years, however, the machines may be dated approximately as follows:

Serial Number	Year	Serial Number	Year
1—10000	1857	100001—115000	1867
10001—20000	1858	115001—130000	1868
20001—30000	1859	130001—145000	1869
30001—40000	1860	145001—160000	1870
40001—50000	1861	160001—190127	1871
50001—60000	1862	190128—223766	1872
60001—70000	1863	223767—239647	1873
70001—80000	1864	239648—253357	1874
80001—90000	1865	253358—267879	1875
90001—100000	1866	267880—279637	1876

Although the Willcox and Gibbs company is still in existence, for the past several decades the company has limited itself to the production of specialized manufacturing machines rather than family machines. (Smithsonian photo 58986.)

Figure 132.—ILLUSTRATION from *Knights American Mechanical Dictionary*, vol. 3, p. 2122. The 68 sewing-machine stitches in use by 1882 are as follows:

SINGLE THREAD
1. Running stitch.
2. Back stitch.
3. Fast stitch.
4. Chainstitch.
5. Coiled-loop chainstitch.
6. Knitted-loop chainstitch.
7. Knotted-loop chainstitch.
8. Loop enchained by second alternate stitch.
9. Each loop locks and enchains alternate loops.
10. Staple stitch (for waxed threads only).

TWO THREADS
11. Double-needle chainstitch.
12. Double-thread chainstitch (one needle).
13. Double-looped chainstitch.
14. Chain with interlocking thread.
15. Under-thread through its own loop.
16. Two needles penetrate fabric from opposite sides.
17. Two needles working from the same side.
18. Double interlocking loop.
19. Lockstitch.
20. Twist in needle thread.
21. Double twist in needle thread.
22. Twist in shuttle thread.
23. Double twist in shuttle thread.
24. Knot stitch, shuttle thread knotted at every stitch.
25. Knot stitch, shuttle thread knotted at every other stitch.
26. Knot stitch, shuttle thread through the needle thread loop and knotted around the loop.
27. Shuttle thread pulled to the surface and interlocked with succeeding stitch to form an embroidery stitch.
28. Wire-lock stitch, thread locked in place with wire.

THREE THREADS
29. Two shuttles, each locking alternate loops.
30. Double loop with interlocking third thread.
31. Two shuttle threads, both locking each loop.
32. Two shuttle threads intertwining and locking each loop.
33. Single thread; loop of needle thread drawn up over the edge and locked by needle at its next descent.
34. Two threads; loops of needle thread, above and below, extend to the edge of the fabric, and are locked by shuttle thread.
35. Two threads; needle penetrates back from edge, its loop passed to and interlocked by the needle at its next descent over the edge, and this second needle-loop locked by shuttle thread.
36. Two threads; shuttle thread drawn up over the edge of the fabric to the line of the needle thread.
37. Two threads; needle loop through the fabric locked by needle loop over the edge and second loop locked by second thread.
38. Two threads; edge of fabric covered by shuttle thread.
39. Three threads; third thread laid around the stitch at the edge of the fabric.

ORNAMENTAL STITCHES
40. Zigzag; single thread chainstitch (4).
41. Zigzag; two-thread lockstitch (19).
42. Zigzag; two-thread chainstitch (13).
43. Zigzag; chain stitch with interlocking thread (14).
44. Zigzag; double loop with interlocking third thread (30).
45. Zigzag; running stitch (1).
46. Zigzag; two needles and shuttle.
47. Zigzag; variation of 46.
48–52. Zigzag stitches for sewing straw braid.
53–62. Straight straw-braid stitches.
63–67. Special embroidery stitches.
68. Saddler's stitch.

In the *Sewing Machine News*, vol. 3, no. 5, p. 12 (1881), there were listed a number of then "defunct" machines and companies. Among these are many well-known names and little-known names for which at least one additional reference can be found. There are some, however, for which this is the only reference to date. These are: Blanchard, Babcock, Banner, Brown Rotary, Cottage, Cole, Duplex, Economist, Erie, Gutman, Hill, Hancock & Bennett, Jenks, Lockmar, La Favorite, Learned, Leggett, McCoy, McCardy, Medallion, McArthur & Co., Monopoly, Moreau, Mack, Niagra, New Cannaan, Orphean, Pride-of-the-West, Seamen & Guiness, Surprise, Stackpole, Shanks, Stanford, Troy, Utica, United States Family, Weaver, Wagner, and Williams. Some of these names may have been a "special" name given to machines manufactured by one of the known companies, but at least a few are names of machines manufactured for a very short time prior to 1881 about which we would like to know more.

III. Chronological List of U.S. Sewing-Machine Patent Models in the Smithsonian Collections

There are more than seven hundred sewing-machine patent models and a similar number of attachment models in the Smithsonian collections. Most of these machines were received in 1926 when the Patent Office disposed of its collection of hundreds of thousands of models. Prior to 1880, models had been required with the patent application; although the requirement was discontinued that year, patentees continued to furnish models for another decade or so. All models prior to 1836 were lost in a Patent Office fire of that year, but since the sewing-machine patent history dates from the 1840s, most of the historically important ones of this subject have been preserved.

These models form a valuable part of the record of the invention, supplementing the drawings and the text of the written specifications. The early sewing-machine models were made to order, either by the inventor or a commissioned model maker. As soon as sewing machines were produced commercially, it was less expensive for the patentee to use a commercial machine of the period, to which he added his change or improvement, than to have a complete model constructed to order. Some of the commercial machines used in this way are the only examples known to be in existence, and as such, are of more interest in establishing the history of the manufactured machine than for the minor patented changes.

During the period of the "Sewing Machine Combination," many patentees attempted to invent and patent "the different machine." This was either a radical change in style or an attempt to produce a far less-expensive type of machine. These machines were not always put into commercial production, but the patent models give an indication of the extent to which some inventors went to simplify or vary the mechanics of machine sewing.

The following is a list of those sewing-machine patent models in the Smithsonian Institution collections:

Patentee	Date	Patent Number
Greenough, John J.	Feb. 21, 1842	2,466
Bean, Benjamin W.	March 4, 1843	2,982
Corliss, George H.	Dec. 27, 1843	3,389
Howe, Elias, Jr.	Sept. 10, 1846	4,750
Bachelder, John	May 8, 1849	6,439
Wilson, Allen B.	Nov. 12, 1850	7,776
Robinson, Frederick R.	Dec. 10, 1850	7,824
Grover & Baker	Feb. 11, 1851	7,931
Singer, Isaac M.	Aug. 12, 1851	8,294
Wilson, Allen B.	Aug. 12, 1851	8,296
Wilson, Allen B.	June 15, 1852	9,041
Miller, Charles	July 20, 1852	9,139
Avery, Otis	Oct. 19, 1852	9,338
Hodgkins, G.	Nov. 2, 1852	9,365
Bradeen, J. G.	Nov. 2, 1852	9,380
Bates, W. G.	Feb. 22, 1853	9,592
Thompson, T. C.	March 29, 1853	9,641
Wickersham, W.	April 19, 1853	9,679
Johnson, W. H.	March 7, 1854	10,597
Harrison, J., Jr.	April 11, 1854	10,763
Avery, Otis	May 9, 1854	10,880
Singer, Isaac	May 30, 1854	10,975
Hunt, Walter	June 27, 1854	11,161
Roper, S. H.	Aug. 15, 1854	11,531
Shaw, P.	Sept. 12, 1854	11,680
Ambler, D. C.	Nov. 1, 1854	11,884
Robertson, T. J. W.	Nov. 28, 1854	12,015
Lyon, W.	Dec. 12, 1854	12,066
Stedman, G. W.	Dec. 12, 1854	12,074
Ward, D. T.	Jan. 2, 1855	12,146
Conant, J. S.	Jan. 16, 1855	12,233
Smith, H. B.	Jan. 16, 1855	12,247
Singer, I. M.	Feb. 6, 1855	12,364
Stedman, G. W.	March 20, 1855	12,573
Stedman, G. W.	May 1, 1855	12,798

Patentee	Date	Patent Number	Patentee	Date	Patent Number
Chilcott, J., and Scrimgeour, J.	March 15, 1855	12,856	Raymond, Charles	March 9, 1858	19,612
			Hendrick, Joseph E.	March 16, 1858	19,660
Durgin, Charles A.	May 22, 1855	12,902	Parker, Sidney	March 16, 1858	19,662
Bond, J., Jr.	May 22, 1855	12,939	Gray, Joshua	March 16, 1858	19,665
Singer, Isaac	June 12, 1855	13,065	Coates, F. S.	March 23, 1858	19,684
Harrison, J., Jr.	Oct. 2, 1855	13,616	Clark, David W.	March 23, 1858	19,732
Singer, I. M.	Oct. 9, 1855	13,661	Reynolds, O. S.	March 30, 1858	19,793
Singer, I. M.	Oct. 9, 1855	13,662	Bartholf, Abraham	April 6, 1858	19,823
Langdon, L. W.	Oct. 30, 1855	13,727	Savage, E.	April 6, 1858	19,876
Stedman, G. W.	Nov. 27, 1855	13,856	Atwood, J. E., J. C., and O.	April 13, 1858	19,903
Swingle, A.	Feb. 5, 1856	14,207			
Watson, Wm. C.	March 11, 1856	14,433	Bosworth, Chas. F.	April 20, 1858	19,979
Singer, I. M.	March 18, 1856	14,475	Clark, David W.	June 8, 1858	20,481
Grover, W. O.	May 27, 1856	14,956	Herron, A. C.	June 15, 1858	20,557
Blodgett, S. C.	Aug. 5, 1856	15,469	Johnson, A. F.	June 22, 1858	20,686
Roper, S. H.	Nov. 4, 1856	16,026	Barnes, W. T.	June 29, 1858	20,688
Singer, Isaac M.	Nov. 4, 1856	16,030	Smith, E. H.	June 29, 1858	20,739
Gibbs, James E. A.	Dec. 16, 1856	16,234	West, H. B., and Willson, H. F.	June 29, 1858	20,753
Jennings, L.	Dec. 16, 1856	16,237			
Johnson, A. F.	Jan. 13, 1857	16,387	Miller, W.	June 29, 1858	20,763
Gibbs, J. E. A.	Jan. 20, 1857	16,434	Blake, Lyman R.	July 6, 1858	20,775
Howe, Elias, Jr.	Jan. 20, 1857	16,436	Carpenter, Lunan	July 27, 1858	20,990
Alexander, Elisa	Feb. 3, 1857	16,518	Moore, Charles	July 27, 1858	21,015
Gray, Joshua	Feb. 3, 1857	16,566	Smith, E. H.	Aug. 3, 1858	21,089
Belcher, C. D.	March 3, 1857	16,710	Wheeler and Carpenter	Aug. 3, 1858	21,100
Pratt, S. F.	March 3, 1857	16,745	Gibbs, J. E. A.	Aug. 10, 1858	21,129
Nettleton & Raymond	April 14, 1857	17,049	Uhlinger, W. P.	Aug. 17, 1858	21,224
Gibbs, J. E. A.	June 2, 1857	17,427	Clark, David W.	Aug. 31, 1858	21,322
Harris, Daniel	June 9, 1857	17,508	Blodgett, S. C.	Sept. 7, 1858	21,465
Harris, Daniel	June 16, 1857	17,571	Hubbard, G. W.	Sept. 14, 1858	21,537
Sage, William	June 30, 1857	17,717	Hendrick, J. E.	Oct. 5, 1858	21,722
Lathbury, E. T.	July 7, 1857	17,744	Gibbs, J. E. A.	Oct. 12, 1858	21,751
Wickersham, W.	Aug. 25, 1857	18,068	Sangster, Amos. W.	Oct. 26, 1858	21,929
Wickersham, W.	Aug. 25, 1857	18,069	Avery, O. and Z. W.	Nov. 9, 1858	22,007
Behn, Henry	Aug. 25, 1857	18,071	Spencer and Lamb	Nov. 23, 1858	22,137
Nettleton, Wm. H., and Raymond, Charles	Oct. 6, 1857	18,350	Perry, James	Nov. 23, 1858	22,148
			Burnet and Broderick	Nov. 30, 1858	22,160
Roper, S. H.	Oct. 27, 1857	18,522	Hook, Albert H.	Nov. 30, 1858	22,179
Fetter, George	Dec. 1, 1857	18,793	Raymond, Charles	Nov. 30, 1858	22,220
Watson, W. C.	Dec. 8, 1857	18,834	Bishop, H. H.	Dec. 7, 1858	22,226
Behn, H.	Dec. 15, 1857	18,880	Pratt, S. F.	Dec. 7, 1858	22,240
Hubbard, George W.	Dec. 22, 1857	18,904	Atwood, J. E.	Dec. 14, 1858	22,273
Lazelle, W. H.	Dec. 22, 1857	18,915	Fosket, W. A., and Savage, Elliot	Jan. 25, 1859	22,719
Clark, David W.	Jan. 5, 1858	19,015			
Fetter, George	Jan. 5, 1858	19,059	Snyder, W.	Feb. 15, 1859	22,987
Clark, David W.	Jan. 12, 1858	19,072	Clark, D. W.	May 3, 1859	23,823
Clark, David W.	Jan. 19, 1858	19,129	Boyd, A. H.	May 17, 1859	24,003
Dimmock, Martial, and Rixford, Nathan	Jan. 19, 1858	19,135	Gray, Joshua	May 17, 1859	24,022
			Hook, Albert H.	May 17, 1859	24,027
Boyd, A. H.	Jan. 19, 1858	19,171	Spencer, James C.	May 17, 1859	24,061
Angell, Benjamin J.	Feb. 9, 1858	19,285	Carhart, Peter S.	May 24, 1859	24,098
Clark, David W.	Feb. 23, 1858	19,409	McCurdy, J. S.	June 14, 1859	24,395

Patentee	Date	Patent Number	Patentee	Date	Patent Number
Goodwyn, H. H.	June 21, 1859	24,455	Gibbs, J. E. A.	June 26, 1860	28,851
Grout, William	July 5, 1859	24,629	McCurdy, J. S.	July 3, 1860	28,993
Hensel, George	July 12, 1859	24,737	Mueller, H.	July 3, 1860	28,996
Parker, Sidney	July 12, 1859	24,780	Sutton, Wm. A.	July 17, 1860	29,202
Hall, William	July 26, 1859	24,870	Hicks, W. C.	July 24, 1860	29,268
Hayden, H. W.	Aug. 2, 1859	24,937	Tracy, D.	Sept. 11, 1860	30,012
Kelsey, D.	Aug. 2, 1859	24,939	Washburn, T. S.	Sept. 11, 1860	30,031
Emswiler, J. B.	Aug. 9, 1859	25,002	Arnold, G. B., and A.	Sept. 25, 1860	30,112
Farr, C. N.	Aug. 9, 1859	25,004	Leavitt, Rufus	Nov. 13, 1860	30,634
Harrison, James, Jr.	Aug. 9, 1859	25,013	Payne, R. S.	Nov. 13, 1860	30,641
Tapley, G. S.	Aug. 9, 1859	25,059	Heyer, Frederick	Nov. 27, 1860	30,731
Barnes, W. T.	Aug. 16, 1859	25,084	Hardie, J. W.	Dec. 4, 1860	30,854
Booth, Ezekial	Aug. 16, 1859	25,087	Earle, T.	Jan. 22, 1861	31,156
Hinkley, J.	Aug. 23, 1859	25,231	Bruen, J. T.	Jan. 22, 1861	31,208
Harrison, James, Jr.	Aug. 30, 1859	25,262	Smith, J. M.	Feb. 5, 1861	31,334
Buell, J. S.	Sept. 13, 1859	25,381	Smith, L. H.	Feb. 12, 1861	31,411
Vogel, Kasimir	Oct. 4, 1859	25,692	Rice, Quartus	Feb. 12, 1861	31,429
Woodward, F. G.	Oct. 11, 1859	25,782	Rose, I. M.	March 5, 1861	31,628
Barrett, O. D.	Oct. 11, 1859	25,785	Ross, Noble G.	March 26, 1861	31,829
Barnes, William T.	Oct. 25, 1859	25,876	Boyd, A. H.	April 2, 1861	31,864
Sawyer, Irwin, and Alsop, T.	Oct. 25, 1859	25,918	Mallary, G. H.	April 2, 1861	31,897
Budlong, William G.	Nov. 1, 1859	25,946	Shaw, H. L.	April 9, 1861	32,007
Fosket, William A., and Savage, E.	Nov. 1, 1859	25,963	Burr, Theodore	April 9, 1861	32,023
Hicks, W. C.	Nov. 8, 1859	26,035	Jones, William, and Haughian, P.	May 14, 1861	32,297
Scofield, C.	Nov. 8, 1859	26,059	Wilder, M. G.	May 14, 1861	32,323
Pearson, William	Nov. 22, 1859	26,201	Smith, Lewis H.	May 21, 1861	32,385
McCurdy, James S.	Nov. 22, 1859	26,234	Stoakes, J. W.	May 28, 1861	32,456
Clark, Edwin	Dec. 6, 1859	26,336	Fuller, William M.	June 4, 1861	32,496
Dickinson, C. W.	Dec. 6, 1859	26,346	Norton, B. F.	July 9, 1861	32,782
Miller, Charles	Dec. 13, 1859	26,462	Raymond, C.	July 9, 1861	32,785
Rowe, Jas.	Dec. 27, 1859	26,638	Raymond, Charles	July 30, 1861	32,925
Johnson, A. F.	Jan. 24, 1860	26,948	Case, G. F.	Aug. 13, 1861	33,029
Thomson, J.	Feb. 7, 1860	27,082	Hodgkins, C.	Aug. 20, 1861	33,085
Juengst, George	Feb. 14, 1860	27,132	Marble, F. E.	Oct. 8, 1861	33,439
Davis, Job A.	Feb. 21, 1860	27,208	Mann, Charles	Oct. 22, 1861	33,556
Gibbs, James E. A.	Feb. 21, 1860	27,214	Grover, W. O.	Nov. 26, 1861	33,778
Rowe, James	Feb. 21, 1860	27,260	Hendrickson, E. M.	Feb. 4, 1862	34,330
Dopp, H. W.	Feb. 28, 1860	27,279	Derocquigny, A. C. F., Gance, D., and Hanzo, L.	March 25, 1862	34,748
Paine, A. R.	March 6, 1860	27,412			
Smalley, J.	March 20, 1860	27,577	Thompson, R.	April 8, 1862	34,926
Newlove, T.	April 3, 1860	27,761	Smith, John C.	April 15, 1862	34,988
McCurdy, J. S.	May 1, 1860	28,097	Palmer, Aaron	May 13, 1862	35,252
Arnold, G. B.	May 8, 1860	28,139	Hall, W. S.	Aug. 5, 1862	36,084
Bean, E. E.	May 8, 1860	28,144	McCurdy, James S.	Aug. 19, 1862	36,256
Holly, Birdsill	May 8, 1860	28,176	Grover, W. O.	Sept. 9, 1862	36,405
Chamberlain, J. N.	May 29, 1860	28,452	Wilkins, J. N.	Sept. 30, 1862	36,591
Ruddick, H.	May 29, 1860	28,538	Humphrey, D. W. G.	Oct. 7, 1862	36,617
Scofield, Chas., and Rice, Clarke	June 5, 1860	28,610	House, H. A., and J. A.	Nov. 11, 1862	36,932
			Crossby, C. O., and Kellogg, H.	Dec. 2, 1862	37,033
Smith, Wilson H.	June 19, 1860	28,785			
Rose, I. M.	June 19, 1860	28,814	Shaw, A. B.	Dec. 16, 1862	37,202

Patentee	Date	Patent Number	Patentee	Date	Patent Number
Pipo, John A.	Jan. 27, 1863	37,550	Rehfuss, George	Nov. 21, 1865	51,086
Hollowell, J. G.	Feb. 10, 1863	37,624	Eickemeyer, Rudolf	Feb. 20, 1866	52,698
Howe, A. B.	March 17, 1863	37,913	Hanlon, John	Feb. 27, 1866	52,847
Weitling, W.	March 17, 1863	37,931	McCurdy, J. S.	April 3, 1866	53,743
Shaw & Clark	April 21, 1863	38,246	Bartram, W. B.	May 15, 1866	54,670
Baldwin, Cyrus W.	April 28, 1863	38,276	Bartram, W. B.	May 15, 1866	54,671
Grote, F. W.	May 5, 1863	38,447	Goodspeed, G. N.	May 15, 1866	54,816
Palmer, C. H.	May 5, 1863	38,450	Hayes, J.	May 22, 1866	55,029
Mack, W. A.	May 19, 1863	38,592	McCloskey, John	June 19, 1866	55,688
Bosworth, C. F.	June 9, 1863	38,807	House, J. A. and H. A.	June 26, 1866	55,865
McCurdy, J. S.	June 16, 1863	38,931	Tucker, Joseph C.	July 24, 1866	56,641
Langdon, Leander W.	July 14, 1863	39,256	Warth, Albin	July 24, 1866	56,646
House, J. A., and H.A.	Aug. 4, 1863	39,442–39,445	Destouy, A.	July 31, 1866	56,729
(4 patents on 1 machine)			Schwalback, M.	July 31, 1866	56,805
Tracy and Hobbs	Sept. 15, 1863	40,000	Cately, William H.	Aug. 7, 1866	56,902
Wagener, Jeptha A.	Oct. 13, 1863	40,296	Piper, D. B.	Aug. 7, 1866	56,990
Rehfuss, G.	Oct. 13, 1863	40,311	Leyden, Austin	Aug. 14, 1866	57,157
Lathrop, Lebbeus W., and	Oct. 27, 1863	40,446	Clements, James M.	Aug. 21, 1866	57,451
de Sanno, Wm. P.			Davis, Job A.	Oct. 9, 1866	58,614
Heyer, W. D.	Nov. 17, 1863	40,622	Rodier, Peter	Nov. 13, 1866	59,659
Simmons, A. G., and	March 1, 1864	41,790	Duchemin, Wm.	Nov. 13, 1866	59,715
Scofield, C.			Kilbourn, E. E.	Nov. 20, 1866	59,746
Guinness, W. S.	March 15, 1864	41,916	Reed, T. K.	Dec. 4, 1866	60,241
Willcox, Charles H.	March 22, 1864	42,036	Singer, I. M.	Dec. 11, 1866	60,433
(4 patents on 1 machine)	Aug. 9, 1864	43,819	Bartram, W. B.	Jan. 1, 1867	60,669
	Sept. 27, 1864	44,490	Rehfuss, G.	Jan. 8, 1867	61,102
	Sept. 27, 1864	44,491	Singer, Isaac	Jan. 15, 1867	61,270
Sibley, J. J.	March 29, 1864	42,117	Cajar, Emil	Feb. 5, 1867	61,711
Thompson, R.	April 19, 1864	42,449	Craige, E. H.	Feb. 19, 1867	62,186
McKay & Blake	May 24, 1864	42,916	Reed, T. K.	Feb. 19, 1867	62,287
Chittenden, H. H.	June 28, 1864	43,289	Bartram, W. B.	March 5, 1867	62,520
Hall, Luther	July 5, 1864	43,404	Fuller, H. W.	March 19, 1867	63,033
Planer, Louis	Aug. 23, 1864	43,927	Stannard, M.	April 23, 1867	64,184
Atwater, B.	Sept. 6, 1864	44,063	Craige, E. H.	Aug. 13, 1867	67,635
Dale, John D.	Oct. 11, 1864	44,686	Doll, Arnold	Sept. 3, 1867	68,420
Gritzner, M. C.	Oct. 18, 1864	44,720	Bruen, L. B.	Sept. 17, 1867	68,839
Smith, DeWitt C.	Dec. 20, 1864	45,528	Hodgkins, C.	Oct. 8, 1867	69,666
Weitling, W.	Jan. 3, 1865	45,777	Baker, G. W.	Oct. 29, 1867	70,152
Cadwell, C.	Jan. 24, 1865	45,972	Cadwell, Caleb	Nov. 19, 1867	71,131
Bartlett, J. W.	Jan. 31, 1865	46,064	Fanning, J.	Dec. 31, 1867	72,829
McCurdy, James S.	Feb. 7, 1865	46,303	Warth, Albin	Jan. 7, 1868	73,064
Lamb, Thomas, and	Aug. 15, 1865	49,421	Rehfuss, George	Jan. 7, 1868	73,119
Allen, John			Cornely, E.	Jan. 28, 1868	73,696
Humphrey, D. W. G.	Aug. 29, 1865	49,627	Blake, L. R.	Feb. 11, 1868	74,289
Tarbox, John N.	Sept. 5, 1865	49,803	Fales, J. F.	Feb. 11, 1868	74,328
Crosby, C. O.	Oct. 3, 1865	50,225	Jencks, G. L.	Feb. 18, 1868	74,694
Cajar, E.	Oct. 3, 1865	50,299	Clark, Edwin E.	Feb. 25, 1868	74,751
Hart, William	Oct. 17, 1865	50,469	Halbert, A. W.	March 31, 1868	76,076
Hecht, A.	Oct. 17, 1865	50,473	Gritzner, M. C.	April 7, 1868	76,323
Emerson, John	Nov. 14, 1865	50,989	Bartlett, Joseph W.	April 7, 1868	76,385
Keats, John, and	Nov. 14, 1865	50,995	Waterbury, Enos	June 16, 1868	79,037
Clark, Wm. S.			Cole, W. H.	June 30, 1868	79,447

Patentee	Date	Patent Number	Patentee	Date	Patent Number
Lamson, Henry P.	July 7, 1868	79,579	Brown, F. H.	April 26, 1870	102,366
French, S.	July 28, 1868	80,345	Howard E., and	May 31, 1870	103,745
Stein, M. J.	Sept. 8, 1868	81,956	Jackson, W. H.		
Hancock, H. J.	Oct. 27, 1868	83,492	Bartram, W. B.	June 14, 1870	104,247
Bartram, W. B.	Nov. 3, 1868	83,592	Henriksen, H. P.	June 21, 1870	104,590
Benedict, C. P.	Nov. 3, 1868	83,596	Martine, Charles F.	June 21, 1870	104,612
Bonnaz, A.	Nov. 10, 1868	83,909	Nasch, Isidor	June 21, 1870	104,630
Bonnaz, A.	Nov. 10, 1868	83,910	Hall, L.	July 12, 1870	105,329
Elliott, F.	Jan. 19, 1869	85,918	Lyon, Lucius	July 26, 1870	105,820
Canfield, F. P.	Jan. 19, 1869	86,057	Bennor, Joseph	Aug. 9, 1870	106,249
Arnold B.	Jan. 26, 1869	86,121	Barnes, M. M.	Aug. 16, 1870	106,307
Jones, John	Jan. 26, 1869	86,163	Leslie, Arthur M.	Oct. 18, 1870	108,492
Russell, W. W.	Feb. 9, 1869	86,695	Rayer, William A., and	Nov. 1, 1870	108,827
Eldridge, G. W.	March 2, 1869	87,331	Lincoln, Wm. S.		
House, J. A. and H. A.	March 2, 1869	87,338	Landfear, Wm. R.	Nov. 22, 1870	109,427
Gird, E. D.	March 9, 1869	87,559	Parham, Charles	Nov. 22, 1870	109,443
Carpenter, William	March 9, 1869	87,633	Lamb, I. W.	Nov. 29, 1870	109,632
Dunbar, C. F.	March 30, 1869	88,282	Moreau, Eugene	Jan. 3, 1871	110,669
McLean, J. N.	March 30, 1869	88,499	Robinson, Charles E.	Jan. 3, 1871	110,790
Billings, C. E.	April 6, 1869	88,603	Goodyear, Charles, Jr.	Jan. 24, 1871	111,197
Winter, Wm.	April 13, 1869	88,936	Stevens, G., and Hendy, J.	Jan. 31, 1871	111,488
Tittman, A.	April 20, 1869	89,093	Carpenter, Mary P.	Feb. 21, 1871	112,016
Swartwout, H. L.	April 27, 1869	89,357	Hancock, Henry J.	Feb. 21, 1871	112,033
Lyons, Lucius	April 27, 1869	89,489	Sidenberg, W.	March 14, 1871	112,745
Crosby, C. O.	May 25, 1869	90,507	Chase, M.	April 11, 1871	113,498
Gutmann, J.	May 25, 1869	90,528	Stein, M. J.	April 11, 1871	113,593
Duchemin, William	June 8, 1869	91,101	Tate, Wm. J.	April 11, 1871	113,704
Adams, John Q.	July 6, 1869	92,138	House, J. A. and H. A.	May 2, 1871	114,294
Bond, Joseph, Jr.	Aug. 10, 1869	93,588	Sidenberg, W.	May 23, 1871	115,117
Hoffman, Geo. W.	Aug. 24, 1869	94,112	Beuttels, Charles	May 23, 1871	115,155
Brown, John H.	Aug. 31, 1869	94,389	Thompson, G.	May 23, 1871	115,255
Heery, Luke	Sept. 14, 1869	94,740	Willcox and Carleton	June 27, 1871	116,521
Gray, Joshua	Oct. 5, 1869	95,581	(3 patents on 1 machine)		116,522
Smith, E. H.	Oct. 26, 1869	96,160			116,523
Page, Chas.	Nov. 2, 1869	96,343	Willcox and Carleton	July 4, 1871	116,783
Lyon, Lucius	Nov. 9, 1869	96,713	Goodyear, Charles, Jr.	July 11, 1871	116,947
Clever, P. J.	Nov. 16, 1869	96,886	Necker, Carl	July 18, 1871	117,101
Mills, Daniel	Nov. 16, 1869	96,944	Pitt, James; Joseph;	July 18, 1871	117,203
Woodruff, Geo. B., and	Nov. 16, 1869	97,014	Edward; and Wm.		
Browning, Geo.			Jones, John T.	Aug. 1, 1871	117,640
Keith, Jeremiah	Dec. 7, 1869	97,518	West, E. P.	Aug. 1, 1871	117,708
Hurtu, Auguste J., and	Dec. 21, 1869	98,064	Jones, Solomon	Aug. 29, 1871	118,537
Hautin, Victor J.			(2 patents on 1 machine)		118,538
Lamb, Thomas	Dec. 28, 1869	98,390	Lamb, Thomas	Sept. 5, 1871	118,728
Rudolph, B.	Feb. 1, 1870	99,481	Bosworth, C. F.	Jan. 9, 1872	122,555
Porter, Alonzo	Feb. 8, 1870	99,704	Smyth, D. M.	Jan. 9, 1872	122,673
Smith, W. T.	Feb. 8, 1870	99,743	Fish, Warren L.	Feb. 13, 1872	123,625
Meyers, N.	Feb. 15, 1870	99,783	Palmer, C. H.	March 19, 1872	124,694
Grover, W. O.	Feb. 22, 1870	100,139	Baker, G. W.	April 9, 1872	125,374
Spoehr, F.	April 12, 1870	101,779	Gordon and Kinert	April 16, 1872	125,807
Kendall, George F.	April 12, 1870	101,887	Howard, C. W.	April 23, 1872	126,056
Cooney, W.	April 26, 1870	102,226	(second machine)		126,057

Patentee	Date	Patent Number	Patentee	Date	Patent Number
Smyth, D. M.	May 14, 1872	126,845	Richardson, E. F.	Dec. 16, 1873	145,687
Beckwith, W. G.	May 21, 1872	126,921	Weber, Theo. A.	Dec. 23, 1873	145,823
Bouscay, Eloi, Jr.	May 28, 1872	127,145	Scribner, Benjamin, Jr.	Jan. 13, 1874	146,483
Braundbeck, E.	June 11, 1872	127,675	Black, Samuel S.	Jan. 20, 1874	146,642
Heidenthal, W.	June 11, 1872	127,765	Taylor, F. B.	Jan. 20, 1874	146,721
Cleminshaw, S.	June 25, 1872	128,363	Richardson, Everett P.	Jan. 27, 1874	146,948
Wardwell, S. W., Jr.	July 2, 1872	128,684	Muir, William	Feb. 3, 1874	147,152
Springer, W. A.	July 9, 1872	128,919	Goodes, E. A.	Feb. 10, 1874	147,387
Fanning, John	July 16, 1872	129,013	Springer, Wm. A.	Feb. 10, 1874	147,441
Parks, Volney	July 30, 1872	129,981	True, C. B.	March 10, 1874	148,336
Baker, G. W.	July 30, 1872	130,005	Wardwell, S. W., Jr.	March 10, 1874	148,339
Smyth, D. M.	Aug. 6, 1872	130,324	Shorey, Samuel W.	March 17, 1874	148,765
McClure, A. T.	Aug. 13, 1872	130,385	Smith, James H.	March 24, 1874	148,902
Ashe, Robert	Aug. 20, 1872	130,555	Horr, Addison D.	April 21, 1874	149,862
Bartram, W. B.	Aug. 20, 1872	130,557	Page, Chas.	May 5, 1874	150,479
West, Elliot P.	Aug. 20, 1872	130,674	Crane, Thomas	May 5, 1874	150,532
Happe, J., and Newman, W.	Aug. 20, 1872	130,715	Buhr, J.	May 26, 1874	151,272
			Smyth, D. M.	June 9, 1874	151,801
Hinds, Jesse L.	Sept. 10, 1872	131,166	Wensley, James	June 16, 1874	152,055
Brown, F. H.	Oct. 1, 1872	131,735	Dinsmore, A. S., and Carter, John T.	Jan. 30, 1874	152,618
Beckwith, W. G.	Nov. 26, 1872	133,351			
Turner, S. S.	Dec. 3, 1872	133,553	Speirs, J.	July 7, 1874	152,813
Chandler, R.	Dec. 10, 1872	133,757	Brewer, A. G.	July 14, 1874	152,894
Venner, O.	Dec. 10, 1872	133,814	Baglin, Wm.	Aug. 18, 1874	154,113
Duchemin, W.	Jan. 21, 1873	135,032	Howard, E. L.	Aug. 25, 1874	154,485
Sheffield, G. V.	Jan. 21, 1873	135,047	Landfear, Wm. R.	Sept. 22, 1874	155,193
Parham, Charles	Feb. 4, 1873	135,579	Drake, Ellis	Oct. 13, 1874	155,932
Goodes, E. A.	March 11, 1873	136,718	Barney, Samuel C.	Oct. 20, 1874	156,119
Tittman, A.	March 11, 1873	136,792	Moreau, Eugene	Oct. 20, 1874	156,171
Happe, J., and Newman, W.	March 25, 1873	137,199	Huntington, Thomas S.	Dec. 29, 1874	158,214
			Bartlett and Plant	Jan. 26, 1875	159,065
Ragan, Daniel	April 1, 1873	137,321	Garland, H. P.	Feb. 16, 1875	159,812
O'Neil, John	April 8, 1873	137,618	Dinsmore, Alfred S.	March 9, 1875	160,512
Kallmeyer, G.	April 8, 1873	137,689	McCloskey, John	March 30, 1875	161,534
Ross, J. G., and Miller, T. L.	May 13, 1873	138,764	Schmidt, Albert E.	April 27, 1875	162,697
			Darling & Darling	May 25, 1875	163,639
West, Elliott P.	May 13, 1873	138,772	Richardson, Everett P.	July 13, 1875	165,506
Koch and Brass	May 13, 1873	138,898	Whitehill, Robert	July 27, 1875	166,172
Arnold, B.	May 20, 1873	138,981	Weber, Theodore A.	Aug. 3, 1875	166,236
Arnold, B.	May 20, 1873	138,982	Pearson, Wm.	Aug. 17, 1875	166,805
Lathrop, L. W.	May 20, 1873	139,067	Beckwith, William G.	Sept. 7, 1875	167,382
Chandler, Rufus	May 27, 1873	139,368	Hall, John S.	Oct. 11, 1875	168,637
Jones, S. H.	July 8, 1873	140,631	Jones, J. T.	Oct. 26, 1875	169,106
Smyth, D. M.	July 22, 1873	141,088	Garland, H. P.	Oct. 26, 1875	169,163
Wardwell, S. W., Jr.	July 29, 1873	141,245	Wormald & Dobson	Nov. 9, 1875	169,881
Stewart, J., Jr.	July 29, 1873	141,397	Rose, R. M.	Nov. 30, 1875	170,596
Walker, William	July 29, 1873	141,407	Keith, Jeremiah	Dec. 7, 1875	170,741
Blanchard, Helen A.	Aug. 19, 1873	141,987	Keith, T. K.	Dec. 14, 1875	170,955
Springer, W. A.	Aug. 26, 1873	142,290	Leavitte, Albert	Dec. 14, 1875	171,147
Cushman, C. S.	Sept. 2, 1873	142,442	Toll, Charles F.	Dec. 14, 1875	171,193
Porter, D. A.	Nov. 25, 1873	144,864	Keats, Greenwood, & Keats	Dec. 28, 1875	171,622
Koch & Brass	Dec. 2, 1873	145,215			

Patentee	Date	Patent Number	Patentee	Date	Patent Number
Thayer, Augustus	Jan. 11, 1876	172,205	Beck, August	Nov. 6, 1877	196,863
Frese, B.	Jan. 18, 1876	172,308	Keith, T. H.	Nov. 6, 1877	196,909
Pearson, William	Jan. 18, 1876	172,478	Keats, John	Dec. 11, 1877	198,120
Sawyer & Esty	Feb. 29, 1876	174,159	Briggs, Thomas	Jan. 1, 1878	198,790
Porter & Baker	March 14, 1876	174,703	Corey, J. W.	Jan. 8, 1878	198,970
Walker, William	April 11, 1876	176,101	Howard, T. S. L.	Jan. 15, 1878	199,206
Upson, L. A.	April 18, 1876	176,153	Bosworth, C. F.	Jan. 22, 1878	199,500
Witherspoon, S. A.	April 18, 1876	176,211	Dancel, C.	Jan. 29, 1878	199,802
Rice, T. M.	April 25, 1876	176,686	Pearson, M. H.	Feb. 5, 1878	199,991
Murphy, E.	May 2, 1876	176,880	Morrell, Robert W.;	April 23, 1878	202,857
Bradford, E. F., and	May 16, 1876	177,371	Parkinson, Thomas; and		
Pierce, V. R.			Parkinson, Joseph		
Applegate & Webb	May 25, 1876	177,784	Barcellos, D.	April 30, 1878	203,102
Sullivan, John J.	June 27, 1876	179,232	Elderfield, F. D.	June 4, 1878	204,429
Appleton, C. J., and	July 4, 1876	179,440	Heberling, J.	June 4, 1878	204,604
Sibley, J. J.			Beukler, William	June 11, 1878	204,704
Marin, Chas.	July 11, 1876	179,709	Varicas, L.	June 11, 1878	204,864
Gullransen, P. E., and	July 25, 1876	180,225	Stewart, W. T.	July 2, 1878	205,698
Rettinger, J. C.			House, Jas. A.	July 23, 1878	206,239
Butcher, Joseph	Aug. 1, 1876	180,542	Martin, W., Jr.; Dawson,	Aug. 6, 1878	206,743
Jackson, William	Sept. 5, 1876	181,941	D. R.; and Orchar, R.		
Barton, Kate C.	Sept. 12, 1876	182,096	Conklin, N. A.	Aug. 6, 1878	206,774
Eickemeyer, Rudolf	Sept. 12, 1876	182,182	Wollenberg, H., and	Aug. 6, 1878	206,848
Webster, W.	Sept. 12, 1876	182,249	Priesner, J.		
Knoch, C. F.	Oct. 17, 1876	183,400	Young, E. S., and	Aug. 13, 1878	206,992
Cushman, C. S.	Nov. 21, 1876	184,594	Dimond, G. H.		
Harris, David	Dec. 12, 1876	185,228	Hoffman, Clara P., and	Aug. 13, 1878	207,035
Wood, J.	Dec. 26, 1876	185,811	Meyers, Nicholas		
Oram, Henry	Jan. 2, 1877	185,952	Wensley, Jas.	Aug. 20, 1878	207,230
Palmer, Frank L.	Jan. 2, 1877	185,954	Dimond, G. H.	Aug. 27, 1878	207,400
Hall, John S.	Feb. 6, 1877	187,006	Steward, A.	Aug. 27, 1878	207,454
Palmateer, William A.	Feb. 20, 1877	187,479	Wood, Richard G.	Sept. 10, 1878	207,928
Cummins, William G.	Feb. 27, 1877	187,822	McCombs, Geo. F.	Sept. 24, 1878	208,407
Esty, William	Feb. 27, 1877	187,837	Keith, Jeremiah	Oct. 22, 1878	209,126
Leavitt & Drew	Feb. 27, 1877	187,874	Wells, W. W.	Nov. 12, 1878	209,843
Henriksen, H. P.	March 20, 1877	188,515	Bayley, C. H.	Feb. 11, 1879	212,122
McKay, Gordon	March 27, 1877	188,809	Parmenter, Charles O.	Feb. 18, 1879	212,495
Follett, J. L.	April 10, 1877	189,446	Ingalls, N., Jr.	Feb. 25, 1879	212,602
Bond, James, Jr.	April 17, 1877	189,599	Cleminshaw, S.	March 18, 1879	213,391
Jacob, F.	April 24, 1877	190,047	Webb, T., and Heartfield,	March 25, 1879	213,537
Beck, A.	May 1, 1877	190,184	C. H.		
Hallett, H. H.	June 5, 1877	191,584	Borton, Stockton	April 8, 1879	214,089
Randel, William	June 12, 1877	192,008	Henriksen, H. P.	May 20, 1879	215,615
Corbett, E., and	July 3, 1877	192,568	Bland, Henry	June 3, 1879	216,016
Harlow, C. F.			Morrison, T. W.	June 10, 1879	216,289
Brown, F. H.	July 24, 1877	193,477	Bosworth, Charles F.	June 17, 1879	216,504
Melhuish, R. M.	Aug. 28, 1877	194,610	Simmons, Frederick	June 24, 1879	216,902
Atwood, K. C.	Sept. 4, 1877	194,759	Junker, Carl	July 1, 1879	217,112
Macaulay, F. A.	Oct. 9, 1877	195,939	Legat, Désiré Mathurin	Aug. 12, 1879	218,388
Dimond, George H.	Oct. 16, 1877	196,198	Willcox, C. H.	Aug. 12, 1879	218,413
Sedmihradsky, A. J.	Oct. 23, 1877	196,486	Cornely, Emile	Sept. 2, 1879	219,225
Keith, J.	Nov. 6, 1877	196,809			

Patentee	Date	Patent Number	Patentee	Date	Patent Number
Hamm, E.	Sept. 16, 1879	219,578	Hull, E. H.	Feb. 28, 1882	254,217
Tuttle, J. W., and Keith, T. K.	Sept. 16, 1879	219,782	Roberts, William	March 7, 1882	254,696
			Willcox and Borton	March 28, 1882	255,576
Stackpole, G., and Applegate, J. H.	Oct. 7, 1879	220,314	Borton and Willcox	March 28, 1882	255,577
			Borton and Willcox	March 28, 1882	255,580
Otis, S. L.	Oct. 28, 1879	221,093	Borton and Willcox	March 28, 1882	255,581
Bland, H.	Nov. 11, 1879	221,505	Veukler, W.	April 4, 1882	255,916
Bracher, T. W.	Nov. 11, 1879	221,508	Hurtu, A. J.	May 30, 1882	258,761
Snediker, J. F.	Nov. 25, 1879	222,089	Keats, Alphonso	July 11, 1882	260,990
Mooney, J. H.	Dec. 2, 1879	222,298	Ramsden, John W.	Aug. 1, 1882	262,116
Osborne, J. H.	Feb. 3, 1880	224,219	Koch, William	Aug. 8, 1882	262,298
Smith, W. M.	March 2, 1880	225,199	Bigelow, J.	Aug. 29, 1882	263,467
Banks, C. M.	March 23, 1880	225,784	Mills, Daniel	Oct. 10, 1882	265,850
Haberling, J.	May 4, 1880	227,249	Wilkinson, Charles E.	Dec. 19, 1882	269,251
Haberling, J.	May 11, 1880	227,525	Carlisle, W. S.	Jan. 9, 1883	270,540
Wiseman, Edmund	June 8, 1880	228,711	Holden, O. J., and Griswold, L.	Feb. 13, 1883	272,050
Juengst, George	June 15, 1880	228,820			
Morley, J. H.	June 15, 1880	228,918	Cameron, James W.	Feb. 20, 1883	272,527
Curtis, G. H. W.	June 22, 1880	228,985	Miller, L. B., and Diehl, P.	March 20, 1883	274,359
Lipe, C. E.	June 29, 1880	229,322	Ludeke, W.	April 10, 1883	275,506
Miller, L. B., and Diehl, P.	July 6, 1880	229,629	Bolton, J., and Petnz, A. D.	May 8, 1883	277,106
Willcox, C. H.	July 20, 1880	230,212	Blodgett, John W.	June 12, 1883	279,320
Shaw, E.	July 27, 1880	230,580	Haberling, J.	Sept. 4, 1883	284,300
Dinsmore, A. S.	Aug. 17, 1880	231,155	Thimonnier, E., and Vernaz, C.	Oct. 30, 1883	287,592
Thurston, C. H.	Oct. 12, 1880	231,300			
Butcher, J.	Oct. 26, 1880	233,657	Duchemin, William	Nov. 20, 1883	288,929
Smyth, D. M.	Nov. 23, 1880	234,732	Lawrence, G. H.	Dec. 25, 1883	290,895
Hesse, J.	Dec. 7, 1880	235,085	Clever, Peter J.	April 8, 1884	296,529
Kjalman, H. N.	Dec. 21, 1880	235,783	Palmer, John H.	May 6, 1884	298,228
Morley, J. H.	Jan. 4, 1881	236,350	Dowling, James, and Connolly, John	May 27, 1884	299,118
Thomas, J.	Jan. 11, 1881	236,466			
Benson, G.	March 8, 1881	238,556	Boecher, Adam	June 10, 1884	300,199
Green, G. F.	March 8, 1881	238,678	Luedeke, Waldemar	June 17, 1884	300,380
Eickemeyer, Rudolf	March 29, 1881	239,319	VanVechten, Orville R.	July 15, 1884	302,063
Palmer, C. H.	April 26, 1881	240,758	Carr, Wm. H., and Ostrom, F. W.	Aug. 12, 1884	303,361
Campbell, D. H.	May 17, 1881	241,612			
Campbell, Duncan H.	May 17, 1881	241,613			
Leslie, A. M.	May 24, 1881	241,808	Trip, J.	Dec. 2, 1884	308,711
Newell, George F.	June 7, 1881	242,470	Farrar, Arthur	Dec. 30, 1884	309,837
Gritzner, Max C.	June 28, 1881	243,444	Turner, M. G.	Feb. 17, 1885	312,306
Keith, Jeremiah	July 5, 1881	243,710	Mills, D.	March 3, 1885	313,359
Choquette, A. E.	July 12, 1881	244,033	Hurtu, August J.	April 7, 1885	315,037
Mooney, J. H.	July 19, 1881	244,470	Charmbury, Henry	April 28, 1885	316,745
Beardslee, W. F.	Aug. 16, 1881	245,781	Woodward & Keith	April 28, 1885	316,927
Hine, Charlie M.	Aug. 23, 1881	246,136	Walker, William	June 16, 1885	320,099
Willcox, C. H.	Sept. 6, 1881	246,700	Tucker, R. D.	June 23, 1885	320,898
Hoefler, J.	Sept. 13, 1881	246,883	Wheeler and Dial	Oct. 13, 1885	328,165
Woodward, E.	Sept. 20, 1881	247,285	Thomas, Joseph	Nov. 10, 1885	330,170
Richards, Jean E.	Jan. 24, 1882	252,799	Muegge, C. A.	Dec. 8, 1885	332,207
Abbott, W. W.	Jan. 31, 1882	252,984	Diehl, P.	April 13, 1886	339,623
Secor, J. B.	Feb. 14, 1882	253,772	Diehl, P.	Aug. 24, 1886	347,776
Deschamps, O. L.	Feb. 21, 1882	253,915			

Patentee	Date	Patent Number	Patentee	Date	Patent Number
Helwig, Arthur	Oct. 5, 1886	350,364	Stewart, James, Jr.	July 15, 1890	432,449
Miehling, Charles	Nov. 2, 1886	351,992	Dewees, J. W.	July 22, 1890	432,746
Dieterle, H. E.	Nov. 30, 1886	353,542	Powell, Thomas	Dec. 16, 1890	442,695
Walker, William	Dec. 7, 1886	353,720	Fletcher, James H.	Dec. 30, 1890	443,756
Rosenthal, S. A.	Dec. 7, 1886	353,970	Rudolph, Ernst B., deceased,	April 7, 1891	449,927
Temple, John	Feb. 22, 1887	358,088	Boulter, W. E.,		
Gee, W. V.	April 19, 1887	361,406	administrator		
Lingley, John W.	Aug. 16, 1887	368,538	Goodwin, Julius C.	April 21, 1891	450,793
Boppel, Jacob	Jan. 29, 1889	396,979	Cook, Hugo	June 23, 1891	454,610
Webster, William	April 30, 1889	402,497	Bowyer, J. T.	June 23, 1891	454,708
Osterhout and Hallenbeck	May 7, 1889	402,610	Willcox, C. H., and	April 5, 1892	472,094
Bennett and Dowling	Aug. 27, 1889	409,728	Borton, S.		
Hine, Charles M.	Jan. 28, 1890	420,382	Legg and Weston	May 17, 1892	474,840
Wheeler, Nathaniel	Feb. 4, 1890	420,847	Kern, Ferdinand	July 19, 1892	479,369
Hallenbeck, J. P.	April 8, 1890	425,422	Jackson, Francis	May 1, 1894	519,064
Lisle, Myron C.	May 20, 1890	428,171	Charles Abercrombi	June 5, 1892	520,977
Walker and Bennet	May 20, 1890	428,548	Taft, J. C.	Oct. 15, 1895	547,866

IV. 19th-Century Sewing-Machine Leaflets in the Smithsonian Collections

Machine or Manufacturer	Date	Type
American B.H.O. and Sewing Machine	1874	Illustrated, advertising leaflet
Buckeye sewing machine	ca. 1870	Illustrated, directions for using the machine
New Buckeye	ca. 1872	Illustrated, directions for using the machine
Centennial sewing machine	1876	Illustrated, advertising leaflet
Domestic sewing machine	1872	Illustrated, advertising leaflet
Florence sewing machine	1873	Illustrated, advertising leaflet
Florence sewing machine	1878	Illustrated, directions for using the machine
Goodes sewing machine	ca. 1876	Advertising leaflet
Grant Brothers sewing machine	1867	Illustrated, advertising leaflet (Xerox copy)
Grover and Baker sewing machine	1853	Illustrated, advertising leaflet
Grover and Baker sewing machine	ca. 1870	Illustrated, advertising leaflet
Home sewing machine	ca. 1870	Illustrated, advertising leaflet
Howe sewing machine, new "B" machine	1868	Illustrated, instruction booklet
Howe sewing machine	1876	Illustrated, catalog of machines
Independent Noiseless sewing machine	ca. 1874	Illustrated, advertising leaflet
Ladd, Webster sewing machine	1861	Illustrated, advertising leaflet
Little Monitor sewing machine	ca. 1872	Illustrated, advertising leaflet
Remington Family sewing machine	ca. 1874	Illustrated, advertising leaflet
Shaw and Clark sewing machine	1864	Illustrated, advertising leaflet
Singer sewing machine	1871	Illustrated, advertising leaflet
Singer sewing machine	1893	Catalog of machines shown at the Columbian Exposition
Standard Shuttle sewing machine	ca. 1875	Illustrated, advertising leaflet
Ten Dollar Novelty sewing machine	ca. 1870	Illustrated, advertising leaflet
Weed sewing machine	1873	Illustrated, advertising leaflet
Wheeler and Wilson sewing machine	ca. 1869	Illustrated, instruction booklet
Wheeler and Wilson sewing machine	ca. 1870–1875	Illustrated, advertising leaflet
Wheeler and Wilson no. 8 machine	ca. 1878	Illustrated, instruction booklet
Wilson sewing machine	1872	Illustrated, advertising leaflet

V. A Brief History of Cotton Thread

Although Samuel Slater's wife is credited with making the first cotton sewing thread from yarns spun at the Pawtucket, Rhode Island, mill in about 1794, cotton thread did not become a manufactured item at that time. Slater turned all his interests to producing cotton-twist yarns needed for the warps of cotton fabrics. By 1809, however, the agents of Almy and Brown, partners and distributors for Slater, were advertising cotton thread as follows:

> Factory Cotton and Thread Store, No. 26 Court Street opposite Concert Hall. George Connell, Agent for Almy and Brown of Providence and Pawtucket Manufactories, has now for sale from eight to ten thousand weight of yarn, for weaving . . . five hundred pounds cotton thread, in hanks, from No. 12 to 60 of a superior quality and very white.*

Although it was a short hop from the spinning of cotton warps to the twisting of these cotton yarns to form a sewing thread, the general manufacture of cotton thread as an industry did not originate in the United States but rather in Scotland in the early 19th century. Napoleon's blockade, which curtailed Great Britain's importation of silk—needed not only for fabrics but also for making heddle strings for the looms—stimulated the production of cotton thread there. James and Patrick Clark, in desperation,

attempted to substitute cotton for silk in their manufacture of these heddle strings. When they were successful, they considered that if cotton could be used successfully for this purpose it could also be made suitable for sewing thread. In 1812 they built a factory in Paisley, Scotland, which had long been noted for its textile industries. The thread was sold in hanks. About 1820 James' sons, James and John, who were now running J. & J. Clark & Co., began to wind the thread on spools. For this service they charged an extra halfpenny, which was refunded when the empty spool was returned. The thread was usually a three-ply or so-called three-cord thread.

About 1815 James Coats, also of Paisley, started manufacturing thread at Ferguslie, Scotland. His two sons took over the company in 1826 and formed the J. & P. Coats Company. Another brother, Andrew Coats, became the selling agent in the United States about 1840. But the cotton-thread industry was not fully launched.

As reported in an 1853 *Scientific American*, there was "more American thread made ten years ago than there is today."** It was not until the six-cord cabled cotton thread, which was suitable for both machine and hand sewing, was perfected that the industry progressed into full operation.

---

*WILLIAM R. BAGNALL, *Textile Industries of the United States* (Cambridge, Mass., 1893), vol. 1, p. 164.

**Scientific American* (Oct. 22, 1853), vol. 9, no. 6, p. 46.

VI. Biographical Sketches

BARTHELEMY THIMONNIER

The first man known to have put a sewing machine into practical operation, Barthelemy Thimonnier, was a Frenchman of obscure parentage. His father, a textile dyer of Lyon, left that city in 1793 as a result of the Revolution and journeyed with his family to l'Arbresle where Barthelemy was born in August of that year.

The family resources were small, and, although the young Thimonnier was able to begin studies at the Seminaire de Saint-Jean at Lyons, he soon was forced to leave school for financial reasons and return to his home, then at Amplepuis. There he learned the tailoring trade and by 1813 was fairly well established in his own shop.

At that time many of the town's inhabitants were weavers and almost every house possessed one or two looms. The noise of the shuttle echoed from these family workshops. Thimonnier noted the relatively small amount of time needed to weave a fabric compared with the slow painstaking task of sewing a garment by passing the needle in and out for each stitch of each seam. When his mind began to dwell on the idea of producing a machine to do this stitching, another of the town's occupations supplied him with a clue and an additional incentive. This village industry produced a type of embroidery work called *point de chainette*, in which a needle with a small hook was used to form the chainstitch, a popular type of decorative stitch long used in countries all over the world. It was Thimonnier's plan to use this type of hooked needle and produce the stitch by machine, employing it both as a decorative stitch and a seam-forming one.

In 1825 Thimonnier moved to St. Etienne, where he became completely absorbed in the idea of inventing a sewing machine. Ignorant of any of the principles of mechanics, he worked alone and in secret for four years, neglecting his tailoring business to the extent that neighbors looked upon him as peculiar, if not crazy. By 1829 he had not only mastered the mechanical difficulties of bringing his dream to realization, but also had made the acquaintance of the man who helped him to success. Ferrand, of l'Ecole des Mines of Saint-Etienne, became interested in the machine and helped finance Thimonnier through his trials and disappointments. In 1830 Thimonnier received a patent on his machine, which produced the chainstitch by means of a needle shaped like a small crochet hook.

Thimonnier, together with Ferrand and a M. Beaunier, made attempts to introduce his machine in Paris. By 1841 they were successful in having eighty of Thimonnier's machines in use sewing army clothing in a shop in Paris. But the fears of the tailors could not be quieted. The machines were destroyed by an ignorant and

Figure 133.—Barthelemy Thimonnier, 1793–1857. From an engraving in the *Sewing Machine Advance*, November 15, 1880. (Smithsonian photo 10569–A.)

infuriated mob, as had been earlier labor-saving devices such as the Jacquard attachment for the loom and Hargreaves' spinning jenny. Thimonnier was forced to flee to his home in St. Etienne, once more penniless.

Soon after this, Jean Marie Magnin, an engineer from Ville-franche-sur-Saône became interested in Thimonnier's machine and provided the inventor again with financial backing. In 1845 under the name of Thimonnier and Magnin the patent of 1830 was renewed, and under it they organized the first French sewing-machine company. The machines they manufactured could produce 200 stitches per minute.

The Revolution of 1848 curtailed the manufacture and sale of the machines. Thimonnier, remembering his unpleasant experience in 1841, decided to go to England with Magnin, where, on February 8, 1848, they received the English patent for his chainstitch machine. He was also granted United States patent 7,622 on September 20, 1850. This later machine had some advantages over his French machine of 1830, but by this time other inventors had joined the field with machines that were more practical. Magnin entered a sewing machine (which from the description in the catalog must have been Thimonnier's invention) in the Crystal Palace Exhibition in London in 1850, but because it was late in arriving it was overlooked by the judges and not even considered in the competition. Thimonnier died in poverty at Amplepuis on July 5, 1857.

WALTER HUNT

Walter Hunt was born near Martinsburg, New York, on July 29, 1796. Although little is known of Hunt's early childhood, we do learn from the author of his obituary, which appeared in *Scientific American*, July 9, 1860, that even as a child he was more interested in people and what he could do for them than in what he could do to insure his own welfare. He is said to have devoted his life to his friends, frequently giving away his last cent when he did not have enough to provide for himself.

There is no record that Hunt maintained a regular business other than the occupation of inventor. His interests were numerous and varied. He received his first patent on June 26, 1826, for a machine for spinning flax and hemp. During the next 33 years he patented 26 ideas. In addition he sold or dropped several more. His second patent was for a coach alarm, and through the years he also received patents for a variety of things including a knife sharpener, heating stove, ice boat, nail machine, inkwell, fountain pen, safety pin, bottle stopper, sewing machine (1854), paper collars, and a reversible metallic heel.

Figure 134.—WALTER HUNT, 1796–1860. From a daguerreotype owned by his great-grandson, C. N. Hunt. (Smithsonian photo 32066–A.)

ELIAS HOWE, JR.

Elias Howe, Jr., was born on his father's farm in Spencer, Massachusetts, on July 9, 1819. This was one of those barren New England farms with many rock-filled acres. All possible ingenuity was necessary to secure a living. The elder Howe supplemented his farming by having a small gristmill, a sawmill, and also by manufacturing cards for the fast-growing cotton industry of New England. Elias Jr.'s earliest recollections were of the latter. He worked with his brothers and sisters sticking wire teeth into strips of leather to make these cotton cards, but, not being very good at this, his family decided to let him "live out" with a neighboring farmer. (Children were leased in those days; they received their board and keep in exchange for chores they would perform.) After a few years, Elias returned home and worked in his father's mill until he was sixteen. Then, against the wishes of his family, he went to Lowell, Massachusetts. Here, he obtained a

138

learner's place in a machine shop where cotton-spinning machinery was made and repaired.

In 1837, when a financial panic hit the country, Howe lost his job. He then decided to go to Boston, and this marked a turning point in his career. In Boston he met Ari Davis, a maker of mariners' instruments and scientific apparatus. Howe began to work in Davis' shop, a place to which inventors often came to ask advice about their ideas. Davis sometimes helped them, but just as often he shouted at them in anger—he is said to have been one of the noisiest men in Boston. One day Howe overheard his employer bellowing at a man who had brought a knitting machine to the shop to seek Davis' advice. "Why are you wasting your time over a knitting machine?" said Davis, "Take my advice, try something that will pay. Make a sewing machine." "It can't be done," was the reply. "Can't be done?" shouted Davis, "Don't tell me that. Why—I can make a sewing machine myself." "If you do," interrupted the capitalist, "I can make an independent fortune for you." Davis, like most men of many words, often talked of more than he planned to do. He never attempted to invent a sewing machine.

But the loud voices interested Howe, who, it is said, determined then that he would produce a sewing machine and win the fortune that the prosperous-looking man had asserted was waiting for such a deed. A kind of lameness since birth had made physical tasks painful for Howe, and he perhaps felt that this would offer an opportunity to become independent of hard physical work.

After marrying on a journeyman machinist's pay of $9 a week, Howe's health worsened and by 1843 was so bad that he had to stop work for days at a time. His wife was forced to take in sewing to maintain the family. It was the sight of his wife toiling at her stitches together with the pressure of poverty that recalled to Howe his earlier interest in a machine to sew. He decided to make an earnest attempt to invent one. Watching his wife for hours at a time, he tried to visualize a machine that would duplicate the motions of the arm. After many trials, he conceived the idea of using an eye-pointed needle in combination with a shuttle to form a stitch. It is possible that, as some authors state, the solution appeared to him in a dream, a manifestation of the subconscious at work. Others have suggested that he may have learned of Hunt's machine. There is a general similarity in the two, not only in the combination of eye-pointed needle and shuttle but in the overhanging arm and vertical cloth suspension.

After conceiving the idea, whatever his inspiration, Howe determined to devote all of his time to producing a working model of his machine. Elias' father, who had then started a factory for splitting palm leaves in Cambridge, gave him permission to set up a lathe and

Figure 135.—ELIAS HOWE, JR., 1819–1867. From an oil painting in the Smithsonian Institution presented by the inventor's grandson, Elias Howe Stockwell. (Smithsonian photo 622.)

a few tools in the garret of the factory. Elias moved his family to Cambridge. Soon after his arrival, unfortunately, the building burned down, and Howe despaired of finding a place to work. He had a friend, however, in George Fisher, who had just come into a small inheritance, and Howe persuaded him to enter into partnership with him for the development of the machine. Fisher agreed to board Howe and his family, which now included two children, while Howe completed the model. Fisher also agreed to supply $500 for material and tools in exchange for a half interest in a patent if one was obtained.

At long last Howe was able to spend his full time and concentration on building his machine. His family was being fed and had a roof over its head. Within a few months Howe had completed a model and by April 1845 had sewed his first seam (see fig. 14). In July of that year he sewed all the principal seams of two suits of wool clothes, one for George Fisher and one for himself.

Several efforts were made to solicit public interest in the new machine. One was installed in a public hall in Boston, and a tailor was employed to operate it at three times the regular wage. The reception was similar to that of Thimonnier's: crowds came to see the "contraption," but, when Howe tried to interest large clothing

139

establishments in using the machine, the protests of the tailors effectively blocked him. He took his sewing machine to the Quincy Hall Clothing Manufactory and offered to sew up any seams brought to him. Daily he sat in one of the rooms demonstrating his machine, and finally he challenged five of the swiftest seamstresses there to a race. Ten seams of equal length were prepared for stitching. One was given to each of the girls while the remaining five were given to Howe. Howe finished his five a little sooner than the girls each finished one, and his seams were declared the strongest and neatest. (Had any curved or angular work been brought, he could not have stitched it.) Still Howe did not receive a single order. The fear of throwing hand sewers out of work was again expressed, and, in addition, the cost of the machine was said to be too high. When it was estimated that a large shirtmaker would have to buy thirty or forty such machines, the necessary large investment was dismissed as ridiculous.

Howe was not too discouraged. In the meantime, he had finished a second machine for deposit with the patent specifications, as the patent laws then required. The second was a better made machine (fig. 15) and showed several minor changes. As soon as the patent was issued on September 10, 1846, Howe and his partner returned to Cambridge.

Without the inventor's enthusiasm or love of his own invention, George Fisher became thoroughly discouraged. He had boarded Howe and his family for nearly two years, had furnished the money needed to purchase the tools and materials for making the two sewing machines, had met the expense of obtaining the patent and the trip of Howe and himself to Washington; representing in all an outlay of practically $2000. Since no orders for machines had been received from either garment makers or tailors, Fisher did not see the slightest probability of the machine's becoming profitable and regarded his advances of cash as a dead loss.

Howe moved back to his father's house with a plan to look elsewhere for a chance to introduce the machine. Obtaining a loan from his father, he built another machine and sent it to England by his brother Amasa. After many discouraging attempts to interest the British, Amasa met William Thomas, a manufacturer of umbrellas, corsets, and leather goods. Thomas employed many workmen, all of whom stitched by hand, and he immediately saw the possibilities of a sewing machine. He proposed that Howe sell the machine to him for £250 sterling (about $1250). Thomas further proposed to engage the inventor to adapt this machine to the making of corsets, at a salary of £3 a week.

When Amasa Howe returned to Cambridge with the news, Elias was reluctant to accept Thomas' offer but had nothing better in sight. So the brothers sailed for London in February 1847, taking with them Howe's first machine and his patent papers. Thomas later advanced the passage money for Howe's wife and three children so that they could join Howe in England.

At this point, historians disagree on how long Howe was in Thomas' employ and whether he succeeded in adapting the machine to meet Thomas' needs. He was in England long enough, however, to find himself without employment in a strange country, his funds nearly exhausted, and his wife ill. He hoped to profit by the notice that his work had received and began to build another machine. He sent his family home to reduce expenses while he stayed on to finish the machine.

After working on it for three or four months, he was forced to sell it for five pounds and to take a note for that. To collect enough for his passage home, he sold the note for four pounds cash and pawned his precious first machine and his patent papers. He landed in New York in April 1849 with but half a crown in his pocket to show for his labors. A short time after he arrived, he learned that his wife was desperately ill. Only with a loan from his father was he able to reach her side before she died. Friends were found to look after the children, and Elias returned to work as a journeyman machinist.

Howe discovered, much to his surprise, that during his absence in England the sewing machine had become recognized in the United States. Several machines made in Boston had been sold to manufacturers and were in daily operation. Upon investigating them, he felt that they utilized all or part of the invention that he had patented in 1846, and he prepared to secure just compensation for its use. The first thing he did was to regain his first machine and patent papers from the London pawnshop. It was no easy matter for Howe to raise the money, but by summer he had managed. It was sent to London with Anson Burlingame, who redeemed the loans, and by autumn of the same year the precious possessions were back in Howe's hands. Though Howe gained nothing by his English experience, William Thomas by his modest expenditure obtained all rights to the machine for Great Britain. This later proved to be a valuable property.

Howe then began writing letters to those whom he considered patent infringers, requesting them to pay a fee or discontinue the manufacture of sewing machines which incorporated his patented inventions. Some at first were willing to pay the fee, but they were persuaded by the others to stand with them and resist Howe. This action forced Howe to the courts. With his father's aid he began a suit, but soon found that considerably more money than either possessed was necessary for such actions. Howe turned once more to George Fisher, but years of investing money in Howe's machine without any monetary return had cooled him to the idea. Fisher, however, agreed to sell his half interest, and in February 1851 George S. Jackson, Daniel C. Johnson, and William E. Whiting became joint owners with Howe. These men helped Howe to procure witnesses in the furtherance of numerous suits, but more money was needed than they could raise. The following year a Massachusetts man by the name of George W. Bliss was persuaded to advance the money for the heavy legal expenses needed to protect the patent. Bliss did this as a speculation and demanded additional security. Once more Elias' long-suffering parent came to the rescue and mortgaged his farm to get the necessary collateral.

Only one of these suits was prosecuted to a hearing, but this one, relatively unimportant in itself, set the precedent. In it the defense relied on the earlier invention of Walter Hunt to oppose Howe's claims. The defendant succeeded in proving that Hunt invented, perfected, and sold two machines in 1834 and 1835 which contained all the essential devices in Howe's machine of 1846. But Howe showed that the defendant's machine (which was a Blodgett and Lerow) contained some features of Howe's machine which were not in Hunt's. The jury decided the case in favor of Howe. Howe later fought a vigorous battle with Isaac Singer, but after much legal controversy the ultimate decision in that case also was in Howe's favor. The suits and payments to each patent holder for the right to use his idea were choking the sewing-machine industry. Even Howe could not manufacture a practical machine without an infringement. Finally an agreement was reached and a "Combination" was formed by the major patent holders (see pp. 41–42).

In the meantime, eight years of the first term of Howe's patent had expired without producing much revenue. This permitted Howe, upon the death of his partner, George Bliss, to buy Bliss' half interest for a small sum. He became, then, the sole owner of his patent just as it was to bring him a fortune. He obtained a seven-year extension for his patent in 1860 without any difficulty, and in 1867, when he applied for another extension, he stated that he had received $1,185,000 from it. Though he endeavored to show that because of the machine's great value to the public he was entitled to receive at least $150,000,000, the second application was denied.

During the Civil War, Howe enlisted as a private soldier in the 17th Regiment Connecticut Volunteers. He went into the field and served as an enlisted man. On occasion when the Government was pressed for funds to pay its soldiers, he advanced the money necessary to pay his entire regiment.

Howe did not establish a sewing-machine factory until just before his death in 1867. One of his early licensees had been his elder brother, Amasa, who had organized the Howe Sewing Machine Company about 1853. When Elias began manufacturing machines on his own, he sunk into the bedplate of each machine a brass medallion bearing his likeness. Elias gave his company the same name that his elder brother had used. As this had been Amasa's exclusive property for many years, he took the matter to the courts where the decision went against Elias. He then organized the Howe Machine Company and began to manufacture sewing machines. On October 3, 1867, Elias died in Brooklyn, New York, at the home of one of his sons-in-law. The company was then carried on by his two sons-in-law, who were Stockwell brothers. In 1872 the Howe Sewing Machine Company was sold by Amasa's son to the Stockwells' Howe Machine Company, which in turn went out of business in the mid-1880s.

ALLEN BENJAMIN WILSON

Allen B. Wilson was born in the small town of Willett, Cortlandt County, New York, in 1824. At sixteen he was apprenticed to a distant relative, a cabinetmaker. Unfortunate circumstances caused him to leave this employ, and in 1847 Wilson was in Adrian, Michigan, working as a journeyman cabinetmaker. The place and year are important, for it was at this time that he conceived his idea of a sewing machine. Because of the distant location, it is believed that he was not aware of similar efforts being made in New England. Wilson became ill and for many months could not work at his trade. By August 1848 he was able to work again and found employment at Pittsfield, Massachusetts. Resolving to develop his idea of a sewing machine, he worked diligently and by November had made full drawings of all the parts, according to his previous conceptions.

In comparison to the monetary returns received by the inventors Howe and Singer, Wilson himself did not receive as great a monetary reward for his outstanding sewing-machine inventions. Because of his health Wilson retired in 1853, when the stock company was formed, but he received a regular salary and additional money from the patent renewals. Wilson petitioned for a second extension of his patents on April 7, 1874, stating that, due to his early poverty, he had been compelled to sell a half interest in a patent (his first one) for the sum of $200. Also he stated that he had not received more than his expenses during the original fourteen-year term. Wilson also stated that he had received only $137,000 during the first seven-year extension period. These figures were verified by his partner. The petition was read before both Houses of Congress and referred to the

Figure 136.—ALLEN BENJAMIN WILSON, 1824–1888. From a drawing owned by the Singer Mfg. Co. Formerly, the drawing was owned by the Wheeler & Wilson Mfg. Co. (Smithsonian photo 32066.)

Committee on Patents.* There was strong feeling against the extension of the Wilson patents. The New York *Daily Graphic*, December 30, 1874, reported:

> So valuable has been this latter four-motion feed that few or no cloth-sewing machines are now made without it. The joint ownership of this feature of the Wilson patents has served to bind the combination of sewing-machine builders together, and enabled them to defy competition by force of the monopoly. It is this feature which the combination wishes to further monopolize for seven years by act of Congress. The inventor has probably realized millions for his invention. Singer admits that his patents, which are much less important, paid him two millions prior to 1870, since which time he has not been compelled to

The Proceedings and Debates of the 43rd Congress, First Session, 1874 Congressional Record, vol. 2, part 3, petition read to the House by Mr. Creamer on April 7, 1874. In part 4 of the same, Mr. Buckingham read a similar petition to the Senate on May 19, 1874. Both were referred to the Committee on Patents; an extension was not granted.

render an account. The Wilson patents with their extended terms were worth a much larger sum. They have been public property, so far as the feed is concerned, since June 15, 1873, and will remain so if too great a pressure is not brought to bear on Congress for their extension. A monopoly of this feed motion for seven years more would be worth from ten to thirty millions to the owner—and would cost the people four times as much.

Wilson had not made the millions for he only received a small percentage of the renewals' earnings plus his salary from the patents' owner, the Wheeler and Wilson Manufacturing Company.

The Congressional Committee on Patents made an adverse report in 1874 and again in 1875 and 1876, when applications for an extension were continued.

Wilson died on April 29, 1888.

ISAAC MERRITT SINGER

Isaac Singer, whose name is known around the world as a manufacturer of sewing machines, was the eighth child of poor German immigrants. Isaac was born on October 27, 1811, in Pittstown, New York, but most of

Figure 137.—ISAAC MERRITT SINGER, 1811–1875. From a charcoal drawing owned by the Singer Mfg. Co. (Smithsonian photo 32066–B.)

his early life was spent in Oswego. He worked as a mechanic and cabinetmaker, but acquired an interest in the theater. Under the name of Isaac Merritt, he went to Rochester and became an actor. In 1839, during an absence from the theater, he completed his first invention, a mechanical excavator, which he sold for $2000. With the money Singer organized a theatrical troupe of his own, which he called "The Merritt Players." When the group failed in Fredericksburg, Ohio, Singer was stranded for lack of funds.

Forced to find some type of employment, Singer took a job in a Fredericksburg plant that manufactured wooden printers' type. He quickly recognized the need for an improved type-carving machine. After inventing and patenting one, he found no financial support in Fredericksburg and decided to take the machine to New York City. Here, the firm of A. B. Taylor and Co. agreed to furnish the money and give Singer room in its Hague Street factory to build machines. A boiler explosion destroyed the first machine, and Taylor refused to advance more money.

While Singer was with Taylor, George B. Zieber, a bookseller who had seen the type-carving machine, considered its value to publishers. Zieber offered to help Singer and raised $1700 to build another model. In June 1850 the machine was completed. Singer and Zieber took the machine to Boston where they rented display space in the steam-powered workshop of Orson C. Phelps at 19 Harvard Place. Only a few publishers came to look at the machine, and none wanted to buy it.

Singer, contemplating his future, became interested in Phelps' work, manufacturing sewing machines for J. A. Lerow and S. C. Blodgett. Phelps welcomed Singer's interest as the design of the mechanism was faulty and purchasers kept returning the machines for repairs. Singer examined the sewing machine with the eyes of a practical machinist. He criticized the action of the shuttle, which passed around a circle, and the needle bar, which pushed a curved needle horizontally. Singer suggested that the shuttle move to and fro in a straight path and that a straight needle be used vertically. Phelps encouraged Singer to abandon the type-carving machine and turn his energies toward the improvement of the sewing machine. Convinced that he could make his ideas work, Singer sketched a rough draft of his proposed machine, and with the support of Zieber and Phelps the work began.

Singer continued to be active in the sewing-machine business until 1863. He made his home in Paris for a short time and then moved to England. While living at Torquay he conceived the idea of a fabulous Greco-Roman mansion, which he planned to have built at Paignton. Singer called it "The Wigwam." Unfortunately, after all his plans, he did not live to see its completion. Singer died on July 23, 1875, of heart disease at the age of sixty-three.

Bibliography

ADAMS, CHARLES K. Sewing machines. In vol. 7 of *Johnson's universal cyclopaedia*, New York: D. Appleton and Company, 1895.

ALEXANDER, EDWIN P. Sewing, plaiting, and felting machines. Pp. 341–353 in *The Practical Mechanic's Journal's record of the great exhibition, 1862*. 1862.

————. On the sewing machine: Its history and progress. *Journal of the Royal Society of Arts* (April 10, 1863), vol. 2, no. 542, p. 358.

AMERICAN HISTORICAL SOCIETY. *The life and works of George H. Corliss*. (Privately printed for Mary Corliss by the Society.) 1930.

Biography of Isaac M. Singer. *The Atlas*, New York, N.Y., March 20, 1853.

BISHOP, J. LEANDER. In vol. 2 of *A history of American manufactures from 1608 to 1860*, by Bishop; Philadelphia: Edward Young and Company, 1866.

BOLTON, JAMES. Sewing machines, special report on special subjects. In vol. 2 of *Report of the Committee on Awards;* Chicago: World's Columbian Exposition, 1893.

BROCKETT, LINUS P. Sewing machines. In vol. 4 of *Johnson's universal cyclopaedia;* New York: D. Appleton and Company, 1874.

BURR, J. B. *The great industries of the United States*. Hartford: J. B. Burr and Hyde, n.d.

COOK, ROSAMOND C. *Sewing machines*. Peoria, Illinois: Manual Arts Press, 1922.

DEPEW, C. M. 100 years of American commerce. In vol. 2, *American sewing machines* by F. G. Bourne. New York: D. O. Haines, 1895.

DISTRICT OF COLUMBIA CIRCUIT COURT. Case of Hunt vs. Howe. In *Federal Cases*, case 6891, book 12, 1855.

DOOLITTLE, WILLIAM H. *Inventions*. Vol. in Library of modern progress. Philadelphia: Modern Progress Publishing Company, 1905.

DURGIN, CHARLES A. *Digest of patents on sewing machines, Feb. 21, 1842—July 1, 1859*. New York: Livesey Brothers, 1859.

FAIRFIELD, GEORGE A. Report on sewing machines. In vol. 3 of *Report of the commissioners of the United States to the international exhibition held at Vienna, 1873*. 1874.

FISKE, BRADLEY A. *Invention, the master-key to progress*. New York: E. P. Dutton and Company, 1921.

GIFFORD, GEORGE W. [Counsel for Elias Howe, Jr.] *Argument of Gifford in favor of Howe's application for extension of patent*. Washington: U.S. Patent Office, 1860.

GRANGER, JEAN. Thimonnier et la machine à coudre. In vol. 2 of *Les publications techniques*. Paris, 1943. [Source for biographical sketch on Barthelemy Thimonnier.]

GREGORY, GEORGE W. *Machines, etc., used in sewing and making clothing*. Vol. 7 of *Reports and Awards*. Philadelphia International Exhibition, 1876. U.S. Centennial Commission, Washington, D.C., 1880.

HERSBERG, RUDOLPH. *The sewing machine, its history, construction and application.* Transl. Upfield Green. London: E. &. F. N. Spon, 1864.

HOWE, HENRY. *Adventures and achievements of Americans.* Published by the author, 1860.

HOWE MACHINE COMPANY. *The Howe exhibition catalog of sewing machines.* Published by the company for the Philadelphia Centennial Exposition, 1876.

HUBERT, PHILIP G., JR. *Men of achievement: inventors.* New York: Charles Scribner's Sons, 1894.

HUNT, CLINTON N. *Walter Hunt, American inventor.* Published by the author, 1935. [Source for biographical sketch on Walter Hunt.]

INTERNATIONAL EXHIBITION. Sewing and knitting machines. In *Reports of the Juries,* Class VII, Section A. London, 1862.

JOHNSON CLARK AND COMPANY. *Sewing machines for domestic and export trade.* Published by the company, 1883.

KAEMPFFERT, WALDEMAR. A popular history of American invention. *Making clothes by machine,* by John Walker Harrington, vol 2. New York: Charles Scribner's Sons, 1924.

KILGARE, C. B. *Sewing-machines.* Philadelphia: J. B. Lippincott Company, 1892.

KNIGHT, EDWARD H. Vol. 3 of *Knight's American mechanical dictionary.* Boston: Houghton, Mifflin and Company, 1882.

LEWTON, FREDERICK L. The servant in the house: a brief history of the sewing machine. Pp. 559–583 in Annual Report of the Smithsonian Institution (1929); Washington: 1930.

LUTH, ERICH. *Ein Mayener Strumpfwirker, Balthasar Krems, 1760-1813, Erfinder der Nähmaschine.* Hamburg, Germany: Verlag Herbert Luth, 1941.

———. *Josef Madersperger oder der unscheinbore Genius.* Hamburg, Germany: Reichsnerband des Deutschen Nähmaschinenhandels EV, 1933 [?].

Obituary of Isaac M. Singer. *New York Tribune,* July 26, 1875.

MADERSPERGER, JOSEPH. *Beschreibung einer Nähmaschine.* Vienna, ca. 1816.

O'BRIEN, WALTER. Sewing machines. *Textile American,* reprinted March 1931. [A paper read at a meeting of the Leicester Textile Society.]

ORCUTT, REV. SAMUEL. *A history of the old town of Stratford and the city of Bridgeport, Connecticut.* 2 vols. Fairfield County Historical Society, 1886.

PARIS UNIVERSAL EXPOSITION. *Reports of the U.S. Commissioners General Survey,* vol. 1, pp. 293–297. 1867.

PARTON, JAMES. History of the sewing machine. *Atlantic Monthly,* May 1867.

RAYMOND, WILLIAM CHANDLER. *Curiosities of the Patent Office.* Syracuse, N.Y.: published by the author, 1888.

REID, J. A. *A monograph prepared for American industrial education conference and exposition.* Elizabethport, N.J.: Singer Manufacturing Company, 1915.

RENTERS, WILHELM. *Praktisches wissin von der Nähmaschine.* Langensalza, Berlin, and Leipzig: Verlag Julius Beltz, 1933.

Scientific American. Issues of January 27, February 3, and November 24, 1849; July 17, 1852; July 9 and November 5, 1859; January 3 and May 16, 1863; October 19, 1867; and July 25, 1896.

SCOTT, JOHN. *Genius rewarded, or the story of the sewing machine.* New York: J. J. Caulon, 1880.

Sewing Machine Advance. Chicago. Vols. 1–35, no. 5, May 1879 to 1913.

Sewing Machine Journal. September 1879 to December 1884.

Sewing Machine News. New York. Vols. 1–6, 1877–1879; n.s. vols. 1–16, 1879–1893.

Sewing Machine Times. New York. Vols. 1–9, 1882–1890; n.s. vols. 1–38 (nos. 1–725), 1891–April 1924.

STAMBAUGH, JOHN P. *Sewing machines at the 22nd annual exhibition of the Maryland Institute, also a history of the sewing machine.* Hartford, Conn.: Weed Sewing Machine Company, 1872.

Story of sewing machine patents. *New York Tribune,* May 23, 1862.

THOMPSON, HOLLAND. The age of invention. Vol. 37 of the Chronicles of America. New Haven: Yale University Press, 1921.

TOWERS, HENRY M. *Historical sketches relating to Spencer, Massachusetts.* Vol. 1. Spencer, Mass.: W. G. Heffernan, 1901.

URQUHART, J. W. *Sewing machinery.* Vol. 8 in Weale's rudimentary scientific and educational series. London: Crosby Lockwood and Co., 1881.

VAN SLYCK, J. D. *New England manufacturers and manufactories.* Vol. 2. Boston: Van Slyck and Company, 1879. [Source for biographical sketch on Allen Benjamin Wilson, Van Slyck, vol. 2, pp. 672–682.]

WALSO, GEORGE C., JR. *History of Bridgeport and vicinity.* Vol. 2. Bridgeport, Conn.: S. J. Clarke Publishing Company, 1917.

Who invented the sewing machine? *The Galaxy* [article unsigned], vol. 4, August 1867.

Indexes

Geographical Index to Companies listed in Appendix II

CONNECTICUT

Bridgeport
  D. W. Clark, 67
  Jerome B. Secor, 68
  Goodbody Sewing Machine Co., 69
  Howe Machine Co., 69
  Secor Machine Co., 72
  American Hand Sewing Machine Co., 72
  Wheeler & Wilson Mfg. Co., 74
Bristol
  Nettleton & Raymond, 72
  Watson & Wooster, 73
Danbury
  Bartram & Fanton Mfg. Co., 66
Hartford
  Morrison, Wilkinson & Co., 71
  Weed Sewing Machine Co., 74
Meriden
  Fosket and Savage, 68
  Charles Parker Co., 72
  Parker Sewing Machine Co., 72
Middletown
  Victor Sewing Machine Co., 73
Norwich
  Greenman and True Mfg. Co., 69
Waterbury
  Waterbury Co., 73
West Meriden
  Parkers, Snow, Brooks & Co., 70

DISTRICT OF COLUMBIA

Washington
  Post Combination Sewing Machine Co., 72

ILLINOIS

Belleville
  Thomas M. Cochrane Co., 71
  J. H. Drew & Co., 71
Belvidere
  June Mfg. Co., 70
  National Sewing Machine Co., 71
Chicago
  Chicago Sewing Machine Co., 67
  Eldredge Sewing Machine Co., 68
  Free Sewing Machine Co., 69
  Scates, Tryber & Sweetland Mfg., 67
  Sigwalt Sewing Machine Co., 67, 73
  H. B. Goodrich, 69
  June Mfg. Co., 70
Rockford
  Free Sewing Machine Co., 69

MAINE

Biddeford
  Shaw & Clark Sewing Machine Co., 71, 73

MASSACHUSETTS

Boston
  O. Phelps, 66
  J. F. Paul & Co., 66
  Boston Sewing Machine Co., 66
  Bradford & Barber, mfgs., 66
  John P. Bowker, 68
  Empire Sewing Machine Co., 68
  Finkle & Lyon Sewing Machine Co., 68, 73
  Grover and Baker Sewing Machine Co., 69
  Nichols and Bliss, 69
  J. B. Nichols & Co., 69
  Nichols, Leavitt & Co., 69, 70
  N. Hunt & Co., 70
  Hunt and Webster, 70
  Emery, Houghton & Co., 70
  Ladd, Webster & Co., 70
  Leavitt & Co., 70
  Leavitt Sewing Machine Co., 70
  Safford & Williams Makers, 71
  C. A. French, 72
  F. R. Robinson, 72
  Howard & Davis, 72
  I. M. Singer & Co., 73
  Butterfield & Stevens Mfg. Co. 74
  Williams & Orvis Sewing Machine Co., 74
Chicopee Falls
  Shaw & Clark Co., 73
  Chicopee Sewing Machine Co., 73
Florence
  Florence Sewing Machine Co., 67, 68
Foxboro
  Foxboro Rotary Shuttle Co., 68
Lowell
  Aetna Sewing Machine Co., 65
Lynn
  Woolridge, Keene and Moore, 67
Orange
  Gold Medal Sewing Machine Co., 69
  Johnson, Clark & Co., 69, 71, 73
  Grout & White, 71
  New Home Sewing Machine Co., 71
Springfield
  Leader Sewing Machine Co., 70
  Springfield Sewing Machine Co., 73

D. B. Wesson Sewing Machine Co., 74
Winchendon
  J. G. Folsom, 68, 69, 71
  William Grout, 71
Worcester
  Goddard, Rice & Co., 66

MICHIGAN

Detroit
  Decker Mfg. Co., 67
  C. G. Gardner, 69

MISSOURI

St. Louis
  Wardwell Mfg. Co., 73

NEW HAMPSHIRE

Dover
  O. L. Reynolds Manufacturing Co., 71
Marlboro
  Thurston Mfg. Co., 67
Mason Village
  Franklin Sewing Machine Co., 68

NEW JERSEY

Elizabethport
  Singer Mfg. Co. (manufactory, not office), 73
Paterson
  Whitney Sewing Machine Co., 74

NEW YORK

Binghamton
  Independent Sewing Machine Co., 70
Brooklyn
  J. H. Lester, 70
  G. L. Du Laney, 70
Ithaca
  Aiken and Felthousen (patentees), 65
  American Magnetic Sewing Machine Co., 65
  Clinton Brothers, 67
  T. C. Thompson, 73
New York
  Avery Sewing Machine Co., 66
  A. Bartholf, mfg., 66
  Bartholf Sewing Machine Co., 66
  Bartlett Sewing Machine Co., 66
  Barlow & Son, 66
  Beckwith Sewing Machine Co., 66

Alphabetical Index to Patentees listed in Appendix III

Dickinson, C. W., 26346

Diehl, Philipp, 339623, 347776, 347777, 348113

Dieterle, H. E., 353542

Dimmock, Martial and Rixford, Nathan, 19135

Dimond, George H., 196198, 207400

Dinsmore, Alfred S., 160512, 231155

Dinsmore, A. S., and Carter, John T., 152618

Doll, Arnold, 68420

Dopp, H. W., 27279

Dowling, James, and Connolly, John, 299118

Drake, Ellis, 155932

Duchemin, William, 59715, 91101, 135032, 288929

Dunbar, C. F., 88282

Durgin, Charles A., 12902

Earle, T., 31156

Eickemeyer, Rudolf, 52698, 182182, 239319

Elderfield, F. D., 204429

Eldridge, G. W., 87331

Elliott, F., 85918

Emerson, John, 50989

Emswiler, J. B., 25002

Esty, William, 187837

Fales, J. F., 74328

Fanning, John, 72829, 129013

Farr, Chester N., 25004

Farrar, Arthur, 309837

Fetter, George, 18793, 19059

Fish, Warren L., 123625

Fletcher, James N., 443756

Follett, Joseph L., 189446

Fosket, William A., and Savage, Elliot, 22719, 25963

French, Stephen, 80345

Frese, B., 172308

Fuller, H. W., 63033

Fuller, William M., 32496

Garland, H. P., 159812, 169163

Gee, W. V., 361406

Gibbs, James E. A., 16234, 16434, 17427, 21129, 21751, 27214, 28851

Gird, E. D., 87559

Goodes, E. A., 136718, 147387

Goodspeed, G. N., 54816

Goodwin, Julius C., 450793

Goodwyn, H. H., 24455

Goodyear, Charles Jr., 111197, 116947

Gordon, James, and Kinert, William, 125807

Gray, Joshua, 16566, 19665, 24022, 95581

Green, George F., 238678

Greenough, John J., 2466

Gritzner, M. C., 44720, 76323

Gritzner, Max C., 243444

Grote, F. W., 38447

Grout, William, 24629

Grover, William O., 14956, 33778, 36405, 100139

Grover, William O., and Baker, William E., 7931

Guinness, William S., 41916

Gullrandsen, P. E., and Rettinger, J. C., 180225

Gutmann, Julius, 90528

Halbert, A. W., 76076

Hall, William, 24870

Hale, William S., 36084

Hall, Luther, 43404

Hall, L., 105329

Hall, John S., 168637, 187006

Hallenbeck, J. P., 425422

Hallett, H. H., 191584

Hamm, E., 219578

Hancock, Henry J., 83492, 112033

Hanlon, John, 52847

Happe, J., and Newman, W., 130715, 137199

Hardie, J. W., 30854

Harris, Daniel, 17508, 17571

Harris, David, 185228

Harrison, J. Jr., 10763, 13616, 25013, 25262

Hart, William, 50469

Hayes, James, 55029

Hayden, H. W., 24937

Heberling, J., 204604, 227249, 227525, 284300

Hecht, A., 50473

Heery, Luke, 94740

Heidenthal, William, 127765

Helwig, Arthur, 350364

Hendrick, Joseph E., 19660, 21722

Hendrickson, E. M., 34330

Henriksen, H. P., 104590, 188515, 215615

Hensel, George, 24737

Herron, A. C., 20557

Hesse, Joseph, 235085

Heyer, W. D., 40622

Heyer, Frederick, 30731

Hicks, William C., 26035, 29268

Hinkley, Jonas, 25231

Hinds, Jesse L., 131166

Hine, Charles M., 420382

Hine, Charlie M., 246136

Hodgkins, Christopher, 9365, 33085, 69666

Hoefler, J., 246883

Hoffman, George W., 94112

Hoffman, Clara P., and Meyers, Nicholas, 207035

Holden, O. J., and Griswold, L., 272050

Hollowell, J. G., 27624

Holly, Birdsill, 28176

Hook, Albert H., 22179, 24027

Horr, Addison D., 149862

House, James A., 206239

House, James A., and House, Henry A., 36932, 39442, 39445, 55865, 87338, 114294

Howard, C. W., 126056, 126057

Howard, T. S. L., 199206

Howard, E. L., 154485

Howard, E., and Jackson, W. H., 103745

Howe, Amasa Bemis, 37913

Howe, Elias, 4750, 16436

Hubbard, George W., 18904, 21537

Hull, E. H., 254217

Humphrey, D. W. G., 36617, 49627

Hunt, Walter, 11161

Huntington, Thomas S., 158214

Hurtu, Auguste J., 258761, 315037

Hurtu, Auguste J., and Hautin, Victor J., 98064

Ingalls, N., Jr., 212602

Jackson, Francis, 519064

Jackson, William, 181941

Jacob, Frederick, 190047

Jencks, G. L., 74694

Jennings, L., 16237

Johnson, Albert F., 16387, 20686, 26948

Johnson, W. H., 10597

Jones, John T., 86163, 117640, 169106

Jones, Samuel H., 140631

Jones, Solomon, 118537, 118538

Jones, William, and Haughian, P., 32297

Juengst, George, 27132, 228820

Junker, Carl, 217112

Kallmeyer, G., 137689

Keats, Alphonso, 260990

Keats, John, 198120

Keats, John; Greenwood, Arthur; and Keats, Alphonso, 171622

Keats, John, and Clark, Wm. S., 50995
Keith, Jeremiah, 97518, 170741, 196809, 209126, 243710
Keith, T. H., 196909
Keith, T. K., 170955
Kelsey, D., 24939
Kendall, George F., 101887
Kern, Ferdinand, 479369
Kilbourn, E. E., 59746
Kjalman, H. N., 235783
Knoch, Charles F., 183400
Koch, William, 262298
Koch, Friederich, and Brass, Robert, 138898, 145215

Lamb, Isaac W., 109632
Lamb, Thomas, 98390, 118728
Lamb, Thomas, and Allen, John, 49421
Lamson, Henry P., 79579
Landfear, William R., 109427, 155193
Langdon, Leander W., 13727, 39256
Lathbury, E. T., 17744
Lathrop, Lebbeus W., 139067
Lathrop, Lebbeus W., and de Sanno, William P., 40446
Lawrence, G. H., 290895
Lazelle, W. H., 18915
Leavitte, Albert, 171147
Leavitt, Rufus, 30634
Leavitt, Albert, and Drew, Henry L., 187874
Legat, D. M., 218388
Legg, Albert, and Weston, Charles W., 474840
Leslie, A. M., 241808
Leyden, Austin, 57157
Lingley, John W., 368538
Lipe, C. E., 229322
Lisle, Myron C., 428171
Ludeke, Waldemar, 275506, 300380
Lyon, Lucius, 89489, 96713, 105820
Lyon, W., 12066

Macauley, F. A., 195939
Mack, William A., 38592
Mallary, G. H., 31897
Mann, Charles, 33556
Marble, F. E., 33439
Marin, Charles, 179709
Martin, W., Jr.; Dawson, D. R.; and Orchar, R., 206743
Martine, Charles F., 104612
Meyers, Nicholas, 99783
Melhuish, R. M., 194610
McCloskey, John, 55688, 161534
McClure, A. T., 130385

McKay, Gordon, and Blake, Lyman, 42916
McKay, Gordon, 188809
McLean, J. N., 88499
McCombs, George F., 208407
McCurdy, James S., 24395, 26234, 28097, 28993, 36256, 38931, 46303, 53743
Miehling, Charles, 351992
Miller, Charles, 9139, 26462
Miller, Westley, 20763
Miller, Lebbens B., and Diehl, Philipp, 229629, 274359
Mills, Daniel, 96944, 265850, 313359
Mooney, John H., 222298, 244470
Moore, Charles, 21015
Moreau, Eugene, 110669, 156171
Morley, J. H., 228918, 236350
Morrell, R. W.; Parkinson, T.; Parkinson, J., 202857
Morrison, T. W., 216289
Muegge, C. A., 332207
Mueller, H., 28996
Muir, William, 147152
Murphy, E., 176880

Nasch, Isidor, 104630
Necker, Carl, 117101
Nettleton, William H., and Raymond, Charles, 17049, 18350
Newell, George F., 242470
Newlove, Thomas, 27761
Norton, B. F., 32782

O'Niel, John, 137618
Oram, Henry, 185952
Osborne, J. H., 224219
Otis, S. L., 221093
Osterhout, James A., and Hallenbeck, Joseph P., 402610

Page, Charles, 96343, 150479
Paine, A. R., 27412
Palmateer, William A., 187479
Palmer, Aaron, 35252
Palmer, C. H., 38450, 124694, 240758
Palmer, Frank L., 185954
Palmer, John H., 298228
Parham, Charles, 109443, 135579
Parker, Sidney, 19662, 24780
Parks, Volney, 129981
Parmenter, Charles O., 212495
Payne, R. S., 30641
Pearson, M. H., 199991
Pearson, William, 26201
Pearson, William, 166805, 172478
Perry, James, 22148

Piper, D. B., 56990
Pipo, John A., 37550
Pitt, James; Joseph; Edward; and William, 117203
Planer, Louis, 43927
Porter, D. A., 144864
Porter, Alonzo, 99704
Porter, D'Arcy, and Baker, George W. 174703
Powell, Thomas, 442695
Pratt, Samuel F., 16745, 22240

Ragan, Daniel, 137321
Ramsden, John W., 262116
Randel, William, 192008
Rayer, William A., and Lincoln, William S., 108827
Raymond, Charles, 19612, 22220, 32785, 32925
Reed, T. K., 60241, 62287
Rehfuss, George, 40311, 51086, 61102, 73119
Reynolds, O. S., 19793
Rice, T. M., 176686
Rice, Quartus, 31429
Richards, Jean E., 252799
Richardson, E. F., 145687
Richardson, Everett P., 146948, 165506
Roberts, William, 254696
Robertson, T. J. W., 12015
Robinson, Charles E., 110790
Robinson, Frederick R., 7824
Rodier, Peter, 59659
Roper, S. H., 11531, 16026, 18522
Rose, Israel M., 28814, 31628
Rose, Reubin M., 170596
Rosenthal, Sally Adolph, 353970
Ross, Noble G., 31829
Ross, J. G., and Miller, T. L., 138764
Rowe, James, 26638, 27260
Ruddick, H., 28538
Rudolph, Bruno, 99481
Rudolph, Ernst B. (deceased), Boulter, W. E. (administrator), 449927
Russell, W. W., 86695

Sage, William, 17717
Sangster, Amos W., 21929
Savage, Elliott, 19876
Sawyer, Irvin P., and Alsop, T., 25918
Sawyer, Sylvanus, and Esty, William, 174159
Schmidt, Albert E., 162697
Schwalback, M., 56805
Scofield, Charles, 26059

Scofield, Charles, and Rice, Clarke, 28610
Scribner, Benjamin, Jr., 146483
Secor, J. B., 253772
Sedmihradsky, A. J., 196486
Shaw, E., 230580
Shaw, Philander, 11680
Shaw, A. B., 37202
Shaw, Henry L., 32007
Shaw and Clark, 38246
Sheffield, G. V., 135047
Shorey, Samuel W., 148765
Sibley, J. J., 42117
Sidenberg, William, 112745, 115117
Simmons, Frederick, 216902
Simmons, A. G., and Scofield, C., 41790
Singer, Isaac M., 8294, 10975, 12364, 13065, 13661, 13662, 14475, 60433, 61270
Smalley, J., 27577
Smith, DeWitt C., 45528
Smith, E. H., 20739, 21089, 96160
Smith, H. B., 12247
Smith, J. M., 31334
Smith, Lewis H., 31411, 32385
Smith, John C., 34988
Smith, James H., 148902
Smith, Wilson H., 28785
Smith, W. M., 225199
Smith, William T., 99743
Smyth, D. M., 126845, 234732
Snediker, J. F., 222089
Snyder, Watson, 22987
Spencer, James C., 24061
Spencer, James H., and Lamb, Thomas, 22137
Speirs, John, 152813
Spoehr, F., 101779
Springer, William A., 128919, 142290, 147441
Stackpole, G., and Applegate, J. H., 220314
Stannard, M., 64184
Stedman, G. W., 12074, 12573, 12798, 13856
Stein, M. J., 81956, 113593
Stevens, George, and Hendy, Joshua, 111488
Steward, A., 207454

Stewart, James, Jr., 141397, 432449
Stewart, W. T., 205698
Stoakes, J. W., 32456
Sullivan, John J., 179232
Sutton, William A., 29202
Swartwout, H. L., 89357
Swingle, A., 14207

Taft, J. C., 547866
Tapley, G. S., 25059
Tarbox, John N., 49803
Tate, William J., 113704
Taylor, F. B., 146721
Temple, John, 358088
Thayer, Augustus, 172205
Thimonnier, E., and Vernaz, C., 287592
Thomas, Joseph, 330170
Thomas, J., 236466
Thompson, J., 27082
Thompson, T. C., 9641
Thompson, Rosewell, 34926, 42449
Thurston, C. H., 233300
Tittman, Alexander, 89093, 136792
Toll, Charles F., 171193
Tracy, Dwight, 30012
Tracy, Dwight, and Hobbs, George, 40000
Trip, J., 308711
True, Cyrus B., 148336
Tucker, R. D., 320898
Tucker, Joseph C., 56641
Turner, M. G., 312306
Turner, S. S., 133553
Tuttle, J. W., and Keith, T. K., 219782
Thompson, George, 115255

Uhlinger, W. P., 21224
Upson, L. A., 176153

Van Vechten, O. R., 302063
Varicas, L., 204864
Venner, O., 133814
Veukler, W., 255916

Wagener, Jeptha A., 40296
Walker, William, 141407, 176101, 320099, 353720
Walker and Bennet, 428548
Ward, D. T., 12146
Wardwell, Simon W., Jr., 128684, 141245, 148339

Warth, Albin, 56646, 73064
Washburn, T. S., 30031
Waterbury, Enos, 79037
Watson, William C., 14433, 18834
Weber, Theodore A., 145823, 166236
Webb, T., and Heartfield, C. H., 213537
Webster, W., 182249, 402497
Weitling, W., 37931, 45777
Wells, W. W., 209843
Wensley, James, 152055, 207230
West, Elliott P., 117708, 130674, 138772
West, H. B., and Willson, H. F., 20753
Wheeler, Nathaniel, 420847
Wheeler, Nathaniel, and Dial, Wilbur F., 328165
Wheeler, Darius, and Carpenter, Lunan, 21100
Whitehill, Robert, 166172
Wickersham, William, 9679, 18068, 18069
Wilder, M. G., 32323
Wilkins, J. N., 36591
Wilkinson, Charles E., 269251
Willcox, Charles H., 42036, 43819, 44490, 44491, 218413, 230212, 246700
Willcox, C. H., and Borton, S., 472094
Willcox, C. H., and Carleton, C. 116521, 116523, 116783
Wilson, Allen B., 7776, 8296, 9041
Wiseman, Edmund, 228711
Witherspoon, S. A., 176211
Winter, William, 88936
Wollenberg, H., and Priesner, J., 206848
Wood, John, 185811
Wood, Richard G., 207928
Woodruff, George B., and Browning, George, 97014
Woodward, E., 247285
Woodward, F. G., 25782
Woodward, Erastus, and Keith, Thomas K., 316927
Wormald, William, and Dobson, Edmund, 169881

Young, E. S., and Dimond, G. H. 206992

General Index to Chapters 1–4

Cover design shows patent
model 99,743 issued to
W. T. Smith on February 8,
1870. The model is in the
Smithsonian collections.